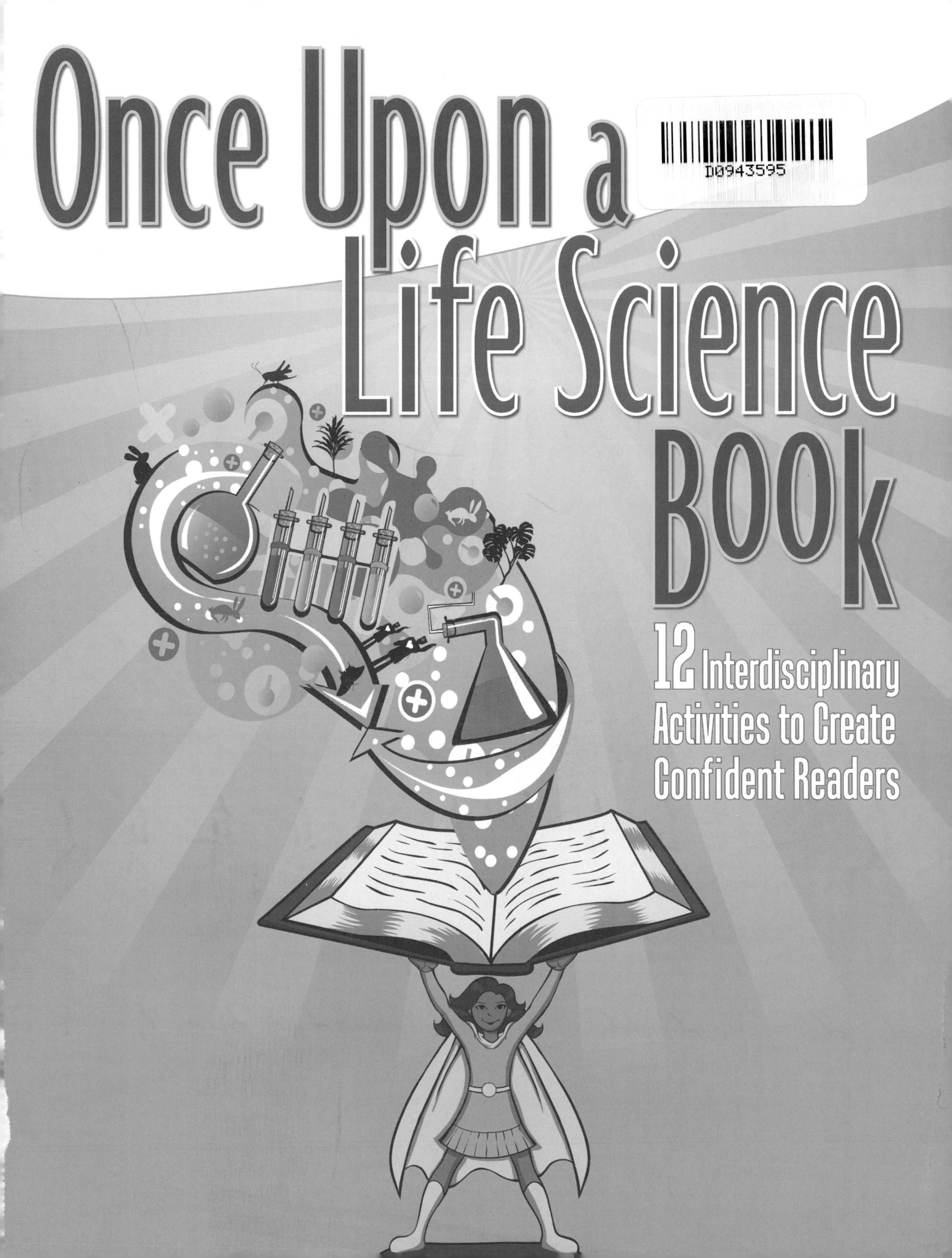
Once Upon a
Life Science
Book
12 Interdisciplinary
Activities to Create
Confident Readers
D0943595

Once Upon a Life Science Book

12 Interdisciplinary Activities to Create Confident Readers

Jodi Wheeler-Toppen

Claire Reinburg, Director
Jennifer Horak, Managing Editor
Andrew Cocke, Senior Editor
Judy Cusick, Senior Editor
Wendy Rubin, Associate Editor
Amy America, Book Acquisitions Coordinator

Art and Design
Will Thomas, Jr., Director—Cover and Interior Design
All superhero art on pages 1, 9, 21, 27, 41, 53, 61, 71, 81, 93, 103, 113 , 123, 135, and 145 courtesy istockphoto.

Printing and Production
Catherine Lorrain, Director

SciLinks
Tyson Brown, Director
Virginie L. Chokouanga, Customer Service and Database Coordinator

National Science Teachers Association
Francis Q. Eberle, PhD, Executive Director
David Beacom, Publisher

12 11 10 4 3 2 1

Library of Congress Cataloging-in-Publication Data
Wheeler-Toppen, Jodi.

Once upon a life science book : 12 interdisciplinary activities to create confident readers / Jodi Lyn Wheeler-Toppen.

p. cm.

Includes bibliographical references and index.

ISBN 978-1-935155-09-6

1. Science--Study and teaching (Secondary)--Activity programs. 2. Reading. 3. Interdisciplinary approach in education. I. Title.

LB1585.W448 2010

507.1--dc22

2009053311

eISBN 978-1-936137-73-2

Table of Contents

Table of Contents

Acknowledgments

For Jon, Natalie, and Zachary

With thanks to

Tammy Moncel
Lac Courte Oreilles Ojibwe Middle School
Hayward, WI

and

Scott Parrish
William Ellis Middle School
Advance, NC

for field testing some of the activities in their classrooms.

And to the following individuals who provided expert reviews of the science content:

Aaron Wheeler
Department of Chemistry
University of Toronto

Peggy Brickman
Division of Biological Sciences
University of Georgia

Norris Armstrong
Division of Biological Sciences
University of Georgia

Lawrence J. Davenport
Department of Biology
Samford University

Acknowledgments

Darold Batzer
Department of Entomology
University of Georgia

David Goldman
National Institute on Alcohol Abuse and Alcoholism
National Institutes of Health

John Heller
Metropolitan Urology Clinic
Minneapolis, MN

Richard Lee
Skidaway Institute of Oceanography
Savannah,GA

Mark Farmer
Department of Cellular Biology
University of Georgia

Jean Pennycook
Education Division
The Penguin Science Project

Wendy Dustman
Department of Microbiology
University of Georgia

J. G. M. "Hans" Thewissen
Department of Anatomy and Neurobiology
Northeastern Ohio Universities College of Medicine

Chapter 1

Getting Started

In faculty meetings, the principal of my school would periodically exhort those of us who were "subject-area teachers" to contribute to the schoolwide emphasis on improving reading. I would return to my classroom and assign pages from the textbook to my students, only to be greeted by moans and groans. After I finally cajoled my groaners into reading, I would ask them questions, but they never seemed to learn much from what they read. Does this sound familiar?

I finally returned to graduate school to find out more about how science teachers could design successful reading lessons for their classes. I learned that there are many parallels between how people learn science and how they learn to become better readers. In fact, several studies have indicated that integrating reading and science can lead to gains in both areas (Fang et al. 2008; Morrow et al. 1997; Romance and Vitale 1992).

This book is for middle and high school life science teachers who are ready to implement successful reading experiences that support science content learning. Each lesson in this book consists of a science activity, a reading about an important life science concept (based on a National Science Education Standard), and an application that asks students to connect what they did with what they read. This book also contains information on teaching reading strategies to help you create a complete reading program in your science classes. First, let's look at how the processes of science and reading are complementary.

Learning Science: The Learning Cycle

Science teachers know that people learn science best when they can anchor their learning in firsthand experiences. This idea has been formalized into the *learning cycle*, a way to organize lessons so that students have a chance to explore a concept before they learn the relevant vocabulary and principles (Lawson 2009). There are actually several versions of the learning cycle; in this book, we will use a three-phase cycle (Karplus and Thier 1967).

Exploration. In this phase, students experience a new concept in a concrete way by exploring a specific example. Explorations range from complex to simple. Students may be asked to design an experiment, attempt to solve a problem, or simply make observations. Examples in this book include designing an experiment to see how mealworms react to different stimuli, dissecting a grasshopper to look for bones, and observing a protist as it swims. The most important aspect of the exploration phase is that students have a chance to develop firsthand knowledge of the concept.

Concept Introduction. In this phase, students learn vocabulary and general principles. For example, if students dissected a grasshopper to look for bones during the exploration phase, they would be introduced to the terms *exoskeleton* and *invertebrate* during the concept introduction phase. They would also learn that the concept they observed during the exploration—that grasshoppers lack bones—is a general principle of invertebrates. In a learning cycle, concept introductions can come from lectures, discussions, readings, or videos. For each cycle in this book, most of the concept introduction is provided through reading.

Concept Application. Finally, students need a chance to apply the new terms and principles themselves in a new situation. The concept application phase of a learning cycle can include doing additional hands-on activities, designing new investigations, making a concept map, or solving a new problem. In this book, the concept application phase for each cycle includes a graphic organizer and writing prompt.

The learning cycle is based in constructivism, a view of learning that holds that students base new knowledge on the understandings they already hold (Lawson 2009). The prior knowledge that students bring to a lesson may be helpful for learning the new material, or students may bring misconceptions that make learning more difficult. The exploration phase can help challenge students' misconceptions so they are ready for the concept introduction.

The exploration also fills in gaps that may be present in a student's prior knowledge. For example, if a student has never looked closely at the roots of a plant, he or she will find it difficult to understand how roots have adapted to absorb water. Observing roots during the exploration phase can provide the necessary background for learning the concepts of plant anatomy.

FIND OUT MORE

If learning cycles are new to you, see Chapters 5 and 6 in

- Lawson, A. 2009. *Teaching inquiry science in middle and secondary schools.* Thousand Oaks, CA: SAGE Publications, Inc.

For an excellent overview of how the learning cycle fits with what we know about how people learn, see the introduction to

- Moyer, R., J. K. Hatchett, and S. A. Everett. 2007. *Teaching science as investigations: Modeling inquiry through learning cycle lessons.* Upper Saddle River, NJ: Pearson Education.

FIND OUT MORE

Read more about prior knowledge and misconceptions in the introduction to

- National Research Council (NRC). 2005. *How students learn: Science in the classroom.* Washington, DC: National Academies Press.

Developing Reading Skills: What Reading Teachers Know

It turns out that prior knowledge is important in reading as well (Rosenblatt 1994). Students often do not understand a text because they lack information that the author of the text assumes they know (Fielding and Pearson 1994). To help students access useful prior knowledge and develop new background that they will need for a text, reading educators use a plan for reading instruction that is very similar to the learning cycle (Robb 2000).

Pre-Reading. During pre-reading, reading teachers ask questions to help students think about what they already know that will be important for understanding the text. For example, if students are going to read about a volcano, the teacher may ask what students know about volcanoes or have them draw a picture of a volcano. Reading teachers call this activating prior knowledge. In some cases, reading teachers need to introduce new background information that students will need for the reading, such as sharing about the time period in which a story is set or showing students an example of an object that plays a key role in the text.

Reading. This step includes the obvious task of reading the passage, but reading teachers may expand it by asking students to use specific strategies to monitor their comprehension as they read.

Post-Reading. After reading, students are led in reflecting on what they learned and applying this knowledge in a new situation. Reading teachers often have students summarize main ideas, create concept maps, or do projects based on what they read.

A Natural Fit

You can see how the work of science teachers and reading teachers fits well together. An exploration can serve as a pre-reading activity by generating background knowledge that supports new learning. It also provides an authentic purpose for reading. No more reading just to answer questions at the end of the chapter—now students can read to answer real questions that they have developed from experiencing science firsthand. For the next phase, reading provides an excellent source for the concept introduction. Reading also models the work of real scientists, who usually read a great deal in the process of developing and interpreting experiments. Finally, both models call for an activity in which students apply their developing knowledge.

One difference in the two models is that reading educators focus more of their attention on what takes place during the actual reading. They know that good readers monitor their comprehension as they read. Reading teachers help students pay attention to whether or not they understand and teach strategies that improve comprehension and memory. Science educators can also teach these strategies, and once again, reading research is on our side—research has shown that students learn reading strategies best if the strategies are incorporated into meaningful reading opportunities (Fielding and Pearson 1994).

Making the Most of This Book

One way to use this book would involve selecting certain lessons that suit your curriculum and using them à la carte. However, this book can also be used as a reading development program by spacing the lessons throughout the year and including the following components:

Strategy Introduction. Each lesson in this book includes a specific reading comprehension strategy that you can introduce before students read. Teaching this strategy does not need to take a great deal of direct instruction; much of the strategy learning will take place during individual practice and small-group interactions. However, explicitly addressing reading strategies can help students learn to take control of their reading (Baker 1991; Spence, Yore, and Williams 1999).

Reading Groups. Working in reading groups can be a powerful tool for improving comprehension (Rosenshine and Meister 1994; Wheeler-Toppen 2006) By working together to read a passage, students can fill in gaps in prior knowledge and model reading strategies for each other (Robb 2000). There are a number of ways to organize successful reading groups, and in the next chapter I will introduce one such procedure that is simple enough to be implemented in most science classrooms.

Journaling. It takes time, practice, and reflection for a new reading strategy to become a stable part of a student's repertoire. Reading teachers often conduct individual reading conferences with their students to reinforce new strategies, monitor progress, and help students reflect on their development as readers (Allen 1995; Schoenbach et al. 1999; Robb 2000). Science teachers rarely have time to implement individual reading conferences in their classes. Therefore, this book includes journal questions to encourage students to internalize the strategies introduced in class. These questions are designed to help students plan ways to use a strategy, practice a strategy by writing an example situation, or evaluate their own use of a strategy. You can increase the value of the journaling activity by periodically responding to entries in your students' journals.

Assessing Student Learning

Assessment is a critical piece of any teaching endeavor. The assessment exercises in this book are based on the idea that assessments should give you feedback on what your students are learning and also serve as learning opportunities for your students (NRC 1996). Therefore, each activity, in addition to being a learning tool, is designed to provide information about how well your students understand the lesson. These assessments fall into four general categories.

The Big Question. The Big Question is a reading comprehension check found at the end of each reading selection. The question can be answered in a fairly straightforward manner from the information in the text but cannot be answered simply by skimming or quoting a section of text. Answers to the Big Question are generally short, and if students work in reading groups, each group may submit just one answer. You should be able to scan the answers relatively quickly to determine if students grasped the main ideas of the text.

Graphic Organizers. Each chapter includes a graphic organizer following the reading selection to help students organize the information they have gathered from the text. Some of the graphic organizers address all of the main points of the text, and others focus on a particular area of importance.

Pulling It Together in Writing. The last activity in each chapter is a writing activity that asks students to integrate what they learned from the exploration and the text. Each activity focuses on a main idea that I will call the Pulling-It-Together Focus Point. The Pulling-It-Together Focus Point will be listed below the writing prompt in each chapter. You can use the rubric in Table 1.1, along with this focus point, to assess your students' responses to the writing prompt. Because these are open-ended questions, they provide an especially good opportunity to watch for misconceptions that your students may have developed.

Claims and Evidence. One science skill highlighted in this book is the ability to use claims and evidence. To this end, six of the explorations ask students to make a claim and support it with evidence. This ability should be assessed as well because what we choose to assess communicates to students what we think is important. Information about assessing the claims and evidence aspects of the explorations can be found in Chapter 3.

Additionally, you will want to assess how students are developing as readers. This is challenging because much of what we teach students to do happens invisibly as they read. However, you can listen to what students say in their reading groups and reading journals. You can also give students the self-evaluation found in Chapter 2 several times during the year to look for growth.

Remember that assessment should affect what you do next with your students (Black 2003). If you find that students are struggling with the Big Question or graphic organizer, you may need to talk about the text with your class. If you encounter misconceptions in the writing responses, you need to address them. If students are struggling with claims and evidence, stop and take a day to try one of the activities from Chapter 3 that will help students develop that skill. Conversely, if students are successful with these activities, you can move on with the confidence that they are ready for new topics.

> **FIND OUT MORE**
> The graphic organizers in this book are expert-generated. There is much evidence that helping students learn to develop their own graphic organizers can increase student learning (Fisher 2002). If you would like to learn more about using student-created graphic organizers, consider the following resource:
> - Berman, S. 2008. *Thinking strategies for science, grades 5–12.* 2nd ed. Thousand Oaks, CA: Corwin Press.

Table 1.1. Rubric for Evaluating Responses to Pulling It Together in Writing Prompts

	Completely	Partially	Incorrectly or Insufficiently
Does the response correctly describe the focus point?			
Is the response supported by details from the reading and/or investigation?			
What misconceptions are present in the response?			

Get Started!

With this book, you have everything you need to boost your students' science and reading skills. Start by learning about the strategies you need for the book in Chapters 2 and 3. Then dive into the 12 content chapters. As you and your students work through these lessons together, you will be able to watch their confidence as readers—and your confidence as a reading educator—grow. So what are you waiting for? Let's get started!

References

Allen, J. 1995. *It's never too late: Leading adolescents to lifelong literacy.* Portsmouth, NH: Heinemann.

Baker, L. 1991. Metacognition, reading, and science education. In *Science learning: Processes and applications,* ed. C. M. Santa and D. E. Alvermann, 2–13. Newark, DE: International Reading Association.

Black, P. 2003. The importance of everyday assessment. In *Everyday assessment in the science classroom*, ed. J. M. Atkin and J. E. Coffey, 1–11. Arlington, VA: NSTA Press.

Fang, Z., L. Lamme, R. Pringle, J. Patrick, J. Sanders, C. Zmach, S. Charbonnet, and M. Henkel. 2008. Integrating reading into middle school science: What we did, found, and learned. *International Journal of Science Education* 30 (15): 2067–2089.

Fielding, L. G., and P. D. Pearson. 1994. Reading comprehension: What works. *Educational Leadership* 51 (5): 62–68.

Fisher, K. M. 2002. Overview of knowledge mapping. In *Mapping biology knowledge*, ed. K. M. Fisher, J. H. Wandersee, and D. E. Moody, 5–24. New York: Kluwer Academic Publishers.

Karplus, R., and H. D. Thier. 1967. *A new look at elementary school science.* Chicago: Rand McNally.

Lawson, A. 2009. *Teaching inquiry science in middle and secondary schools.* Thousand Oaks, CA: SAGE Publications.

Morrow, L. M., M. Pressley, J. K. Smith, and M. Smith. 1997. The effect of a literature-based program integrated into literacy and science instruction with children from diverse backgrounds. *Reading Research Quarterly* 32 (1): 54–76.

National Research Council (NRC). 1996. *National science education standards.* Washington, DC: National Academies Press.

Robb, L. 2000. *Teaching reading in middle school: A strategic approach to reading that improves comprehension and thinking.* New York: Scholastic Professional Books.

Romance, N. R., and M. R. Vitale. 1992. A curriculum strategy that expands time for in-depth elementary science instruction by using science-based reading strategies: Effects of a year-long study in grade four. *Journal of Research in Science Teaching* 29 (6): 545–554.

Rosenblatt, L. 1994. The transactional theory of reading and writing. In *Theoretical models and processes of reading*, ed. R. Ruddell, M. Ruddell, and H. Singers, 1057–1092. 4th ed. Newark, DE: International Reading Association.

Rosenshine, B., and C. Meister. 1994. Reciprocal teaching: A review of the research. *Review of Educational Research* 64 (4): 479–530.

Schoenbach, R., C. Greenleaf, C. Cziko, and L. Hurwitz. 1999. *Reading for understanding: A guide to improving reading in middle and high school classrooms.* New York: Jossey-Bass.

Spence, D. J., L. D. Yore, and R. L. Williams. 1999. The effects of explicit science reading instruction on selected grade 7 students' metacognition and comprehension of specific science text. *Journal of Elementary Science Education* 11 (2): 15–30.

Wheeler-Toppen, J. 2006. Reading as investigation: Using reading to support and extend inquiry in science classrooms. PhD diss., Univ. of Georgia, Athens.

Chapter 2

The Reading Strategies

The belief that reading is essentially a process of saying the words rather than actively constructing meaning from texts is widespread among many students. For instance, one of the students we interviewed looked surprised when he was asked to describe the topic discussed in a section of text he had just read. "I don't know what it was about," he answered, with no sense of irony. "I was busy reading. I wasn't paying attention." (Schoenbach et al. 1999, p. 6)

What Is Reading?

> **TEACHING NOTE**
> Science textbooks are examples of *expository text*, a type of writing that describes or explains a concept. Expository text stands in contrast to *narrative text*, which tells a story. Many science magazine articles use the story of a specific situation to introduce science concepts, so they include both expository and narrative text. I have written some of the articles in this book to reflect textbook-style writing and some to reflect magazine-style writing.

At lunch, my colleagues and I would periodically bemoan how our students "can't read." But what did we really mean by that? Certainly, most of our students, even those scoring well below grade level on reading tests, could pronounce the words on the page of a simple book. Some of them even enjoyed reading novels for fun. But they seemed completely unable to make sense of their science textbooks or other school books.

Part of the problem lay in the way they thought about reading. Like the young man in the example above, they believed that reading consisted of calling out the words on the page. Good readers, they assumed, automatically understood all of those words, and their own failure to do so simply reinforced students' beliefs that they didn't read well. Reading strategies can be important for helping students improve their reading, but students need something more. They need to begin to view reading as an active search for meaning that is within their control. We can change how our students think about reading through the way we talk about reading in our classrooms.

Starting the Conversation

The first step is to create a classroom culture in which students feel safe exploring new ways of thinking about reading. You can begin by simply stating aloud that reading can be difficult, even for good readers. You can share your own stories about encountering words, phrases, or books that were hard for you to understand. Most important, you should make it clear that you will not tolerate students teasing each other about reading struggles.

Next, students need a chance to see what experienced readers *do* as they read. Good readers constantly monitor their comprehension and notice if they do not understand what they read. They often have an ongoing conversation in their head in which they compare what they are reading with what they already know (and sometimes argue with the text if they disagree). When they do not understand, or they find inconsistencies with their prior knowledge, they use problem-solving strategies to make sense of the text or resolve the inconsistencies. All of these things are hidden from someone watching, but as teachers we can make them visible.

Think-Alouds. One way to make the invisible processes of reading visible is to talk about what we are thinking as we read (Kucan and Beck

1997). This is called a *think-aloud*. For example, you might read this section from the reading selection in Chapter 11:

> *I was familiar with how a caterpillar turns into a butterfly. But I had no idea that other animals had similar changes. The blue crab, it turns out, comes out of its egg as this little zoea. It drifts in the currents out at sea, eating algae and hoping to escape the sea trout, croaker, and jellyfish that would love to gobble it up.*

To use this in a think-aloud, you would insert your own thoughts as you read out loud, so it might sound something like this:

> *I was familiar with how a caterpillar turns into a butterfly.* Oh yeah, I remember learning the butterfly life cycle in elementary school. *But I had no idea that other animals had similar changes.* I bet it's about to tell me that a crab changes form, because this article is about crabs. *The blue crab, it turns out, comes out of its egg as this little zoea.* Aha, I was right. *It drifts in the currents out at sea, eating algae and hoping to escape the sea trout, croaker, and jellyfish that would love to gobble it up.* I'm not sure what a croaker is, but I know a trout is a fish, so I'm guessing a croaker is too.

This allows struggling readers to "see" how strong readers approach difficult reading passages. You can use a think-aloud to demonstrate specific strategies to your class or when you are helping a student or small group figure out a confusing passage.

Peer Conversation. Students can also show invisible aspects of reading to each other. In reading groups (see discussion below), students share with each other how they made sense of the text. One student might read the above passage and ask the group, "What's this *z* word?" Another student might answer by saying that it must be something like a caterpillar. A third student might add that the sentence says "A zoea is the crab's stage when it comes out of the egg."

You may be skeptical that your students would be able to have these discussions with each other. You will be surprised. As students get used to working in groups, and as an atmosphere of trust develops, even weak readers become comfortable asking for help and sharing what they think as they read.

Overarching Strategies

Each lesson in this book can be used to introduce students to one or more specific reading strategies. These strategies are important; they represent ways that good readers solve specific reading problems. However, keep in mind that learning specific strategies is not the ultimate goal. We want students to begin to approach reading as an active search for meaning (Loxterman, Beck, and McKeown 1994). The following strategies are only a means to that end.

The first two strategies introduced here, comprehension coding and reading groups, are intended to be used throughout all of the lessons. They address the overarching issues of comprehension monitoring and problem solving.

Comprehension Coding. In comprehension coding, students mark codes to indicate what they are thinking as they read. I recommend introducing the following codes:

! This is important.
✓ I knew that.
x This is different from what I thought.
? I don't understand.

Students do not need to mark every sentence. They can select those for which a code seems appropriate.

Over time, students may develop their own coding systems that meet their particular needs. Indeed, you may notice that you do something similar yourself. You may underline important information you want to remember or jot questions in the margins of books. This strategy is intended to mimic that sort of behavior on the part of good readers and encourage students to monitor their comprehension as they read.

Reading Groups. Working in reading groups can be a powerful tool for improving comprehension (e.g., Rosenshine and Meister 1994; Wheeler-Toppen 2006). There are a number of ways to organize reading groups; however, I recommend the following simple procedure for the activities in this book.

In this procedure, each reading group has three students with specific jobs. The *leader* guides the group through the procedure that is listed on the board or a sheet of paper (see Figure 2.1). The *flag flyer* raises a red flag when the group needs help from the teacher. This flag can be a folded piece

TEACHING NOTE

Some teachers have difficulty using comprehension coding with their students because their school restricts the number of copies they can make. If you are in this position, consider the following ideas:

- Talk to your administrator about the problem. Most administrators are interested in supporting attempts to improve reading. They may be willing to allow you extra copies for this purpose.
- Consider printing a class set of readings and have them laminated or slip them into clear page protectors. Students can code using overhead projector pens and then wipe off their marks for the next class. These class sets can be used for several years.
- To use this strategy with textbooks, you can give students strips cut from sticky notes to mark sentences as they read.

of construction paper that is propped up or a plastic cup that is placed on the desk when help is needed. The *interpreter* reads aloud and then later records the group's answer to the Big Question, a general comprehension check that follows the reading (see Chapter 1 for further explanation).

Figure 2.1. Reading Group Procedure

1. The *interpreter* reads the Big Question aloud. Remember, this is what you are trying to learn!
2. Everyone reads the first section quietly and marks !,✓, x, and ? while reading.
3. The *leader* asks each member of the group to share anything that was confusing (marked ? or x).
4. The group should try to figure out what the confusing word, sentence, or idea means.
5. If the group cannot make sense of the confusing word, sentence, or idea, the *flag flyer* should raise the flag to get help from the teacher.
6. Repeat Steps 1 through 4 for Section 2.
7. Repeat Steps 1 through 4 for Section 3.
8. The group should work together to answer the Big Question. The *interpreter* will write the group's answer to turn in to the teacher.

The group procedure calls for reading passages to be broken into three sections. You will find that the readings in this book are divided into sections by short black lines so they can be used with or without this procedure.

Your role as the teacher is important during this process. Initially, you may need to monitor students closely to ensure that they really do follow the procedure. When students raise red flags, listen to their comprehension difficulty. What have they tried so far? Can you model a strategy for making sense of the text? Is there a piece of background knowledge or a word definition that you need to provide?

Listen in on groups that are not having trouble as well. Encourage students to share their strategies. For example, if one student tells another what a sentence means, ask, "How did you know that?" By participating in the groups with students, you show that even teachers have to think carefully about what they read.

As with any classroom procedure, this one takes practice. For the first few sessions, students will have to focus as much on what to do as on what they are reading. They will also need to see that you are serious about requiring them to follow the procedure. After two or three sessions, however, students will begin to follow the procedure automatically and be able to focus more on content. The time invested in learning the process will be well worth it, as students' reading skills and confidence improve.

Problem-Solving Strategies

The rest of the strategies in this book are designed to help students solve specific comprehension problems or learn something about how science texts are organized. Each chapter will describe how to introduce one of the following strategies that would be appropriate to use while reading the article for that chapter. Before giving the article to your class, read it yourself and identify places the strategy would be useful. This will help you guide your students' reading.

Keep in mind that students need practice to master any strategy. For this reason, it will be important for you to monitor students closely the first time they use a strategy. In this book, several strategies appear in two different lessons to reinforce their use. You may also want to follow up with readings from your textbook or other sources to allow them to practice the strategies further.

Finding the Meaning of New Words. Specialized vocabulary is a key feature of science texts (Fang 2006; Holliday 1991). As students read, they are continually introduced to new terms. Struggling readers often miss the definitions of the words when they are introduced because they don't recognize the cues that a definition is being given.

Most students will have been introduced to the strategy of using context clues to find the meaning of new words. As a general reading strategy, using context clues means looking at the surrounding text to figure out

FIND OUT MORE
For more information on these and other reading strategies, check out the following books:

- Schoenbach, R., C. Greenleaf, C. Cziko, and L. Hurwitz. 1999. *Reading for understanding: A guide to improving reading in middle and high school classrooms.* New York: Jossey-Bass.
- Robb, L. 2003. *Teaching reading in social studies, science, and math.* New York: Scholastic Teaching Resources.

Table 2.1. Common Ways That Texts Introduce New Words

Example	Explanation
Some bacteria are anaerobes. Anaerobes don't need oxygen to live.	The sentence after the term provides a definition.
Bacteria that don't need oxygen to live, called anaerobes, ...	The new term is signaled with the word *called*.
Some bacteria are anaerobes, which means they don't need oxygen to live.	The definition is signaled with the phrase *which means*.
Anaerobes, or bacteria that don't need oxygen to live, ...	The word *or* **after a comma** indicates that the word and phrase mean the same thing.
Some bacteria don't need oxygen to live. These anaerobes ...	This is the trickiest situation. The text doesn't directly say what the word means, but implies it by using the word and the definition close together.

a likely meaning of the unfamiliar word. Although many students know they should "use context clues," they often do not use the strategy successfully. Chapters 5 and 14 help students practice this skill by focusing on some of the most common ways that new definitions are presented in science text (see Table 2.1). Students can learn to look for the clues in the sentence before and after a new word is used for the first time.

Note that sometimes the text does not provide sufficient context clues for students to figure out the meaning of a word, especially for nonscience words or vocabulary that is not the focus of the reading. These words constitute background knowledge that the writer expects students to have already. One advantage to having students in reading groups is that they can help each other with these words. Alternatively, if you can identify words that may cause problems, you can teach the words before giving students the text.

Previewing Diagrams and Illustrations. When students read science books, they often ignore the diagrams and illustrations (Wheeler-Toppen 2006). By doing this, they miss out on some excellent reading support (Holliday 1991). Diagrams and illustrations can clarify points that are hard to explain in words. They may provide useful background knowledge or give examples. Pictures can be especially helpful for struggling readers because they provide hints about what the text will say.

Previewing diagrams and illustrations can serve several functions. When you, as the teacher, lead the preview, you can use it to call attention to important points that the text will make. You can also use it to help students draw on their prior knowledge by having them look for things that they recognize in the pictures. Previewing can also be a useful tool for helping students make predictions before they read. When students begin reading, they will attempt to find out if their predictions were correct. Chapters 6 and 8 take advantage of all of these possibilities. Ultimately, though, we want students to be able to use diagrams and illustrations without teacher guidance. To that end, the most important point to make with your students is that diagrams and illustrations are an important part of science texts. Students should get in the habit of studying diagrams and illustrations, preferably before they read.

Text Signals. Expository text often includes key words that can signal what the reader can expect to find in the next few phrases or sentences (Schoenbach et al. 1999). Some teachers have compared these text signals to traffic signs. When readers see these words or phrases, they should slow

TEACHING NOTE

Sending students to the dictionary, or even the glossary in the back of a book, is not a particularly productive way to help them with unknown words. Dictionary definitions can be more confusing than the original text and include additional words that students do not know. Furthermore, the task of looking up the word interrupts a student's reading process. It is much less disruptive to simply tell students the meaning of a word if it cannot be figured out from the context.

down and notice the information that follows. Chapters 7 and 13 focus on two common groups of signal words. Chapter 7 looks at signals for examples and lists, and Chapter 13 presents signals for comparing and contrasting (see Table 2.2).

Although these are the only signal words that are introduced formally in this book, there are many other types. For example, words such as *following*, *previously*, and *during* indicate that a timeline is being given. The presence of a question in a text generally indicates that an answer will be given. Phrases such as *because of*, *in response to*, and *as a result of* tell the reader that a cause-and-effect relationship is being described. You may find opportunities to introduce these or other text signals in your conversations with students about reading.

Table 2.2. Text Signals Introduced in Chapters 7 and 13

Examples	*for example*, *like*, *including*, *such as*, *to illustrate*, *for instance*, and *e.g.* (stands for the Latin *exempli gratia* and means *for example*)
Lists	Statement with a number (such as "There are 4 main differences." or "Three different approaches were used.") A statement followed by a colon (such as "Use all of your senses: taste, touch, hearing, smell, and sight.")
Comparisons	*in the same way*, *just like*, *just as*, *likewise*, and *also*
Contrasts	*however*, *in contrast*, *on the other hand*, *conversely*, and *whereas* (*but*, *yet*, and *while* sometimes indicate a contrast)

Chunking. Science texts often have a lot of information crammed into each sentence (Fang 2006). Consider the following basic sentence about mosses:

Seedless, nonvascular plants such as mosses have a two-stage life cycle.

To understand the subject of this sentence—seedless, nonvascular plants—the reader has to understand that the sentence is about plants; that these plants do not have seeds; that they are nonvascular; and that because the word *vascular* is probably relatively new to the reader, he or she must also remember what a vascular system is and that *non* means *not*. And this is all before they even get to the main point of the sentence!

The example sentence also includes the phrase *such as mosses*, which is stuck in the middle of the main thought. This type of phrase, which is common in science writing, is called the interruption construction (Fang 2006) because it interrupts the flow of the sentence. If the interruption is long, it can be especially confusing for struggling readers.

Experienced readers intuitively break sentences like the one above into chunks and think about each chunk individually. Struggling readers may try to understand the whole sentence at one time (Schoenbach et al. 1999). The reading strategy of chunking helps students break sentences into separate ideas. Chapters 9 and 12 introduce students to chunking by showing them that they can stop to think as they read, even when there is no period or comma to signal a pause.

Pause, Retell, and Compare. Sometimes students will tell me that they read their book, but they can't remember what they read. They may be working so hard to understand individual sentences that they fail to grasp what the text as a whole covers. Another issue, especially with science, is that the text may contain so much information that it is difficult to remember. Pause, retell, and compare is a strategy to help students remember what they read.

In the pause, retell, and compare strategy, students stop throughout the text to rehearse the information they have just read and then compare it back to the original text. Chapters 10 and 11 provide a modified reading group procedure that includes taking time to pause, retell, and compare. In groups, other students can fill in details that an individual missed. To use this strategy when they are working alone, students should reread the section to see what else they need to remember.

Reading Scientific Names. The final strategy in this book deals with a problem unique to reading about life science—reading scientific names. I have noticed that struggling readers find scientific names challenging. They come across the long, complicated words and announce, "I don't know what that means." If they see several names in one text, they may give up. However, if they can recognize that what they are seeing is a name, and not a word they are supposed to know, they can read through the names without being overwhelmed.

Chapter 15 helps students learn to recognize the form of the scientific name, using cues such as italics, a capitalized first word, and the context. Then they select some strategies for pronouncing the name themselves. Students are often reassured to learn that they can still be successful understanding the text even if they do not pronounce the scientific name correctly.

Using Self-Assessment to Monitor Strategy Development

As discussed in Chapter 1, reading teachers often use individual and small-group conferences to assess how students are developing as readers. Such conferences are beyond the scope of what most science teachers are able to incorporate into their classes. Therefore, you can monitor your students' strategy use and reading confidence by using a periodic self-assessment. To conduct the assessment, have students answer the following questions in their reading journals (adapted from Robb 2000):

- What do you do before reading to get ready to learn?
- While reading, what do you do if you come to a word or section that you do not understand?
- How do you help yourself remember the details of your reading?
- What would you like to do better as a reader?

Then give students a slip of paper with the chart in Table 2.3 to complete and tape into their journals.

Give the assessment first at the beginning of the year, before you start to teach the strategies. Read over your students' answers, as you may want to make changes to your instruction based on what they

Table 2.3. Strategy Use Self-Assessment

How Often Do I … ?	A Lot	Sometimes	Never
Use codes (such as ✓, +, ?, and x) to mark what I'm thinking as I read			
Use the information around a new word to figure out what it means			
Study the diagrams and illustrations before reading			
Use text signals to recognize examples and lists			
Use text signals to recognize comparisons and contrasts			
Chunk difficult sentences into smaller pieces to read			
Recognize scientific names when I come across them			

say. Give the assessment again midyear and at the end of the year, and encourage students to look over their previous responses and note how they have improved.

References

Fang, Z. 2006. The language demands of science reading in middle school. *International Journal of Science Education* 28 (5): 491–520.

Holliday, W. G. 1991. Helping students learn effectively from science text. In *Science learning: Processes and applications,* ed. C. M. Santa and D. E. Alvermann, 38–47. Newark, DE: International Reading Association.

Kucan, L., and I. L. Beck. 1997. Thinking aloud and reading comprehension research: Inquiry, instruction, and social interaction. *Review of Educational Research* 67 (3): 271–299.

Loxterman, J. A., I. L. Beck, and M. G. McKeown. 1994. The effects of thinking aloud during reading on students' comprehension of more or less coherent text. *Reading Research Quarterly* 29 (4): 352–367.

Robb, L. 2000. *Teaching reading in middle school: A strategic approach to reading that improves comprehension and thinking*. New York: Scholastic Professional Books.

Rosenshine, B., and C. Meister. 1994. Reciprocal teaching: A review of the research. *Review of Educational Research* 64 (4): 479–530.

Roth, K. J. 1991. Reading science texts for conceptual change. In *Science learning: Processes and applications,* ed. C. M. Santa and D. E. Alvermann, 48–63. Newark, DE: International Reading Association.

Schoenbach, R., C. Greenleaf, C. Cziko, and L. Hurwitz. 1999. *Reading for understanding: A guide to improving reading in middle and high school classrooms*. New York: Jossey-Bass.

Wheeler-Toppen, J. 2006. Reading as investigation: Using reading to support and extend inquiry in science classrooms. PhD diss., Univ. of Georgia, Athens.

Chapter 3

How Do You Know That?

Helping Students With Claims and Evidence

Many of the activities in this book ask students to make a claim and provide evidence for the claim. Making claims (often called conclusions) and providing evidence are at the heart of the practice of science. A major goal in science education is for students to "develop the abilities necessary to do scientific inquiry" (NRC 1996, p.143), so students need to have opportunities to make claims and support them with evidence. Furthermore, when students write and talk about the claims and evidence they have made, their content learning improves (Norton-Meier et al. 2008).

In academic jargon, making claims and providing evidence is called *argumentation*. But this is not the everyday use of the term *argument*. In everyday language, an argument implies that at least two people are interacting and emotions are running high. In the academic sense, *argumentation* simply refers to the reasoning that someone uses to prove a point.

Scientific Argumentation

In the 1950s, the philosopher Stephen Toulmin identified six major features of scientific arguments, four of which we will examine here: claims, evidence, warrants, and backings (Toulmin 1958). According to Toulmin's model, in a scientific argument evidence is presented to support a claim. Warrants then explain why this evidence provides legitimate support. Backings—essentially, evidence that a warrant is valid—are sometimes used as well.

Consider this everyday example. I came home the other day and found my shoe lying in pieces in the living room and my dog hiding under the bed. Using Toulmin's model, I make the claim that my dog chewed up my shoe, as follows in Figure 3.1 (adapted from an example in Lawson 2003):

Figure 3.1. Using Toulmin's Model to Support a Claim

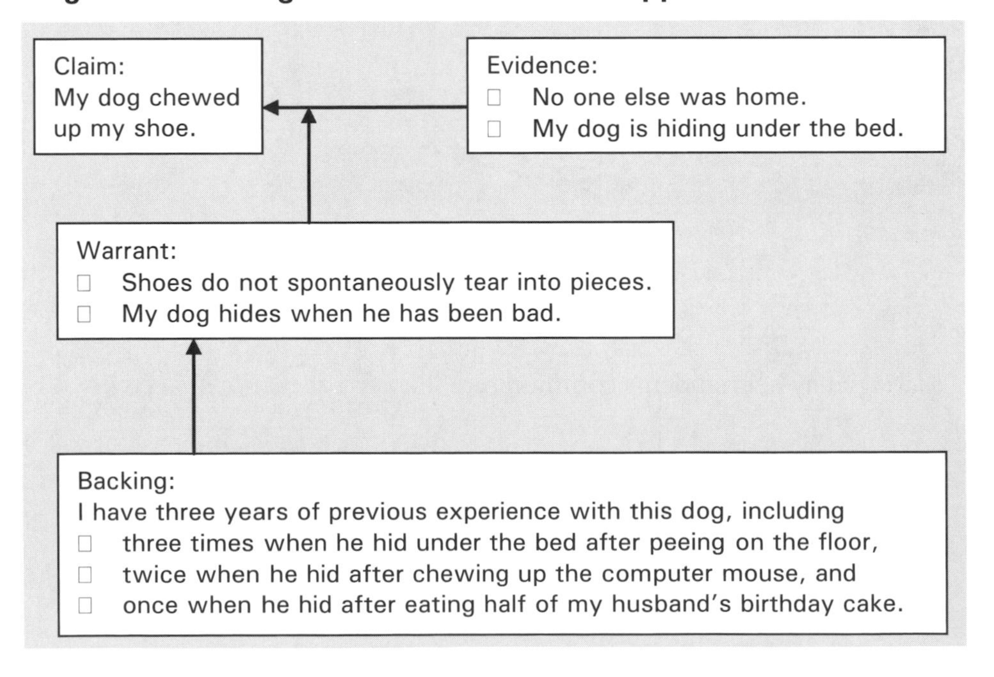

Argument in the Classroom

For the purposes of the middle school classroom, the most important aspect of Toulmin's model is coordinating claims and evidence. Students are surprisingly inexperienced with identifying appropriate evidence to support a scientific claim (Kuhn 1993; Kuhn, Amsel, and O'Loughlin 1998; Zohar 1998). When asked to give evidence to support a claim, students tend to offer reasons why their claim is plausible rather than evidence from data. For example, a student group may have done an experiment in which they let cockroaches choose between light and dark environments. Students may make the claim that cockroaches prefer dark environments. They may even have excellent evidence to support that claim, such as that 9 out of their 10 test subjects moved to a dark environment when given a choice. When asked to provide evidence for the claim, however, students may state something like this:

> *Because cockroaches have those long antennae, maybe they can just feel their way around and don't need light to see.*

Now that's a fine hypothesis, and one that could be tested further, but it is not what scientists would use as evidence to support the claim that cockroaches prefer dark environments. Scientists would provide the data that 9 out of 10 cockroaches chose the dark environment. By helping our students understand what counts as evidence for a scientific claim, we help them understand the nature of science.

Even though you will probably not choose to introduce the terms *warrant* and *backing* to your students, the concepts behind these terms can be useful for you as a teacher. For example, as your students become more comfortable with argumentation, they will begin to challenge their classmates' claims and evidence, often on the grounds that the evidence does not adequately support the claim. You will know that they are challenging the warrant, and drawing a chart similar to Figure A (without the terms *warrant* and *backing*) may help students communicate their critiques more effectively.

Introducing Claims and Evidence

Any simple activity that has students focus on making a claim and supporting it with evidence can be used as a starting point for introducing claims and evidence. I describe two possible approaches here.

1. Who Broke Mrs. Garcia's Bottle of Perfume?

Most students are familiar with thinking about argumentation in legal environments, such as detective work and court trials. One way to introduce claims and evidence is to use this as a starting point.

The story presented at the end of this chapter, *Who Broke Mrs. Garcia's Bottle of Perfume?* (p. 26), describes a simple "crime" and three possible suspects. Have your students read the story and then break into groups. Each group should decide who they think committed the crime. This will be their claim. They should then list specific evidence from the story that supports this claim. Note that the story is intentionally ambiguous so that students can see how different groups interpret the same evidence.

This is a good time to point out that there are different forms of argumentation. Most people are familiar with argumentation by lawyers. Like scientists, lawyers make claims and support them with evidence. However, a lawyer's main interest is to make claims that help his or her client. Counterevidence, or evidence that would hurt a client's claim, is downplayed. In science, however, the goal is to find the claim that best explains all existing evidence, so looking for counterevidence is an important part of making a scientific claim (Lawson 2003).

2. What's in the Box? (adapted from Norton-Meier et al. 2008)

A more active introduction to claims and evidence is to allow students to collect data themselves. You can do this by using mystery boxes. Choose a variety of everyday objects and place them into bags or boxes that students cannot see through. Seal the bags or boxes and give one to each group. Tell students that they should try to determine what is in the bag or box. They may collect evidence in any way they wish as long as they do not open or damage the container. When they believe they know what is inside, they should make a claim and list their evidence.

> ***SAFETY ALERT!***
> Do not use pointed objects, sharps, or other items that can cut or puncture skin.

Groups can then present their claims and evidence to each other. Encourage your students to politely challenge their classmates' claims if they feel the evidence is not strong enough.

Assessing Claims and Evidence

The explorations in Chapters 4, 5, 7, 8, 12, and 14, along with the writing prompt in Chapter 12, ask students to make a claim and support it with evidence. The rubric in Table 3.1 can be used to assess the claims that students make in these activities and also serve as a springboard for a class discussion about what makes an effective claim. If you find that students are struggling with some aspect of claims and evidence, you may wish to revisit the activities in this chapter to introduce claims and evidence to your students.

Table 3.1. Rubric for Assessing Claims and Evidence

	Completely	Partially	Not at All
Claim • Is it the type of claim that an experiment could verify?			
• Does the claim address the question asked in the investigation?			
Evidence • Does the evidence support the claim?			
• Is the evidence sufficient to support the claim—that is, do you have enough information?			
• Is the claim correct based on the data from the investigation?			

References

Kuhn, D. 1993. Science as argument: Implications for teaching and learning scientific thinking. *Science Education* 73 (3): 319–337.

Kuhn, D., E. Amsel, and M. O'Loughlin. 1988. *The development of scientific thinking skills*. San Diego: Harcourt Brace Jovanovich.

Lawson, A. E. 2003. The nature and development of hypothetico-predictive argumentation with implications for science teaching. *International Journal of Science Education* 25 (11): 1387–1408.

National Research Council (NRC). 1996. *National science education standards*. Washington, DC: National Academies Press.

Norton-Meir, L., B. Hand, L. Hockenberry, and K. Wise. 2008. *Questions, claims, and evidence: The important place of argument in children's science writing*. Arlington, VA: NSTA Press.

Toulmin, S. E. 1958. *The uses of argument*. Cambridge: Cambridge University Press.

Zohar, A. 1998. Result or conclusion: Students' differentiation between experimental results and conclusions. *Journal of Biological Education* 32 (1): 53–59.

REMEMBER YOUR CODES
- ! This is important.
- ✓ I knew that.
- X This is different from what I thought.
- ? I don't understand.

Who Broke Mrs. Garcia's Bottle of Perfume?

Someone broke Mrs. Garcia's new bottle of jasmine perfume. She called Marcus, the neighborhood detective, to help her figure out who did it.

"I stepped out for just a minute," said Mrs. Garcia. "When I left to get the mail, my new bottle of perfume was sitting on top of the dresser. And just look at it now, broken all over the floor."

Marcus agreed. It was definitely broken. There were pieces of purple glass everywhere, even in the hallway. And the smell of jasmine was overpowering.

"It had to have been either Ivan or Jessica, the two kids who stay with me in the afternoons," said Mrs. Garcia. "But they both say they didn't do it."

Ivan and Jessica

Marcus went into the den to talk to Ivan and Jessica.

"Jessica has been begging Mrs. Garcia to let her try the perfume," said Ivan. "I bet she did it."

"Did not," said Jessica. "I can't even reach the top of the dresser."

"But I thought I left the step stool in the kitchen," said Mrs. Garcia, "and now it's in the hallway. You could have used that."

"No, I think Ivan did it," said Jessica.

Ivan snorted. "Why would I want some silly perfume?"

"Maybe you were trying to get something else, like the coin jar, when you broke it. Besides, you always lie. You got in trouble for lying last week when you said you didn't eat the extra candy bar."

Mrs. Garcia nodded.

"Well, maybe it was the cat. She's always running around the house," said Ivan.

"When you chase her," said Mrs. Garcia.

More Evidence

Marcus said, "Let me look around for a few minutes. Then I'll tell you what I think."

He started by looking carefully at Ivan and Jessica. He spied a small bit of purple glass stuck in the sole of Jessica's left shoe.

Then he put his nose to work. He sniffed the step stool in the hallway. It didn't smell of jasmine. The cat, however, was another story. He found the little tabby hiding under Mrs. Garcia's bed, and it had a definite jasmine smell.

Marcus headed back to the den, trying to decide what to say to Mrs. Garcia.

If you were Marcus, what would you claim happened? What is your evidence?

Claim:

Evidence:

Chapter 4

A-Maze-ing Worms

Topics

- Scientific method
- Controlling variables

NSES Content Standards (For Grades 5–8, Science as Inquiry)

As a result of activities in grades 5–8, all students should develop

- abilities necessary to do scientific inquiry
- understanding about scientific inquiry (NRC 1996, p. 143)

Reading Strategies

- Comprehension coding
- Reading in groups

Background

This chapter has two main goals. The first goal is to ease students into the reading procedures described in Chapter 1. For that reason, this chapter has two reading passages. "On Your Mark!" allows students to practice using the codes to show what they are thinking as they read. This passage is followed by "A-Maze-ing Worms," which can be used to let students practice the group reading process. If you choose not to have your students work in groups, using two reading passages with this lesson will reinforce that reading is an important part of your science class.

The second goal is to review for students how to design a controlled experiment. Scientists in the life science fields use various types of research to answer questions. They may study a fossil record to confirm or disprove evidence of a hypothesis. They may observe natural systems in action and look for patterns.

Controlled experiments, however, are a cornerstone of life science research. Although most students will have studied this topic, usually called "the" scientific method, in prior classes, many have not mastered it by middle school. In the exploration for this chapter, students will practice designing their own controlled experiments using mealworms.

SciLinks
THE WORLD'S A CLICK AWAY
Topic: Controlled Experiments
Go to: *www.scilinks.org*
Code: LSB001

SAFETY ALERT!
- Remind students to keep water and ice away from electrical devices such as outlets to avoid getting shocked.
- Upon completing work with mealworms, wash hands with soap and water.

Materials

- Mealworms
- Paper plates or containers to use for experiments
- Items to help create a variety of conditions for the mealworms, such as construction paper, plastic wrap, flashlights, paper towels, water, a heating pad, and ice
- Gloves, aprons, indirectly vented chemical splash goggles

Student Pages

- "On Your Mark!"
- "A-Maze-ing Worms"
- A Wormy Question: Design Your Own Controlled Experiment
- What to Do in Your Reading Group
- A-Maze-ing Worms: Diagramming Dr. Wheeler's Experiment

Exploration/Pre-Reading

Because there are two reading passages for this chapter, the exploration is broken into two parts. Before the first reading, students will spend a few minutes observing mealworms and thinking about experiments they might do using the question "How do mealworms respond to ________?" Then they will read "On Your Mark." Next, they will design and conduct their experiments. Finally, they will read "A-Maze-ing Worms" and do the application activities.

The first step in this exploration is to introduce students to mealworms. Give each group of students two or three mealworms and have them make observations for about three minutes. Tell them to note both what mealworms look like and what they do. Students may interact with the mealworms as long as they are careful not to injure them. Explain that mealworms are not actually worms, but a small, black beetle in the larval stage. This period of exploration will give students enough experience with mealworms to be able to design an experiment.

Collect the mealworms so that you will have the students' attention for the following discussion. Tell students that they will design their own experiments to find out something about mealworm behavior. Write "How do mealworms respond to ________?" on the board and have the class brainstorm a list of things that could go in the blank. If they have trouble, you can provide one idea to get them started. Possible responses include (among other things) water, food, a bright light, a dark area, a rough surface, loud noise, or touch.

Introduce the First Reading. Tell students that before they design their experiment, they are going to read a passage that reviews some important things about designing an experiment. Have students read "On Your Mark!" and introduce the idea of using codes to record what they are thinking, as described in the Reading Strategies section.

When they have finished reading "On Your Mark!" place students into lab groups and hand out A Wormy Question. Give the groups a few minutes to select the condition they would like to test, referring them to the list generated by the class if they need help. They should then be ready to design their experiments.

Most students will design an experiment in which the worm is presented with the stimulus for some trials and placed in the same container without the stimulus for others. Other arrangements are possible, however, and students should be allowed to proceed with any reasonable study they design.

If students have difficulty identifying a control group, you could prompt them by asking, "What if the mealworms always move like that, even if you don't have (water, light, or other stimulus)? How could you make sure the mealworm is responding to the stimulus?" Or give the hint, "You will need to compare what the mealworm does when (water, light, or other stimulus) is present with what it does normally." Most students will use a mealworm in an empty container as the control group.

Some students have difficulty deciding what to use as a dependent variable to measure the results of their experiment. Such students may find it helpful if the class generates a list of potential measurements. Possibilities include how far the mealworms crawl, how many times out of 10 the mealworms move toward the stimulus, or how many times in a minute (or other length of time) the mealworms perform any particular movement.

When you have approved each group's design to ensure that it is safe, allow students to gather the materials that they need and conduct their experiments.

Introduce the Second Reading. Tell students they are going to read a second passage that describes an experiment done by a scientist. As they read, they can notice the things he did to make sure his experiment was a fair test.

TEACHING NOTE
Mealworms are great classroom animals because they are inexpensive, easy to rear, and easily available. Most pet stores sell containers of 25 to 30 live mealworms for less than five dollars (ask for them in the reptile section). They can be kept dormant in a refrigerator for several days until you are ready to use them. After the activity, they can be raised in a small container of plain oatmeal to allow students to observe the mealworms' metamorphosis into beetles.
If mealworms are not available in your area, you can substitute any number of other invertebrates, including earthworms, pill or sow bugs (roly-polies), fiddler crabs, or even lab-raised cockroaches.

Reading Strategy 1 (for "On Your Mark!"): Comprehension Coding

To introduce the strategy, use a projector to display a copy of "On Your Mark!" so students can watch as you code. Read the first two paragraphs out loud and model the coding process. For example, you might read the first paragraph and then write a question mark, saying, "This is confusing to me because I was expecting to read about an experiment, not a race." In the second paragraph, you might place an exclamation point after the term *controlled experiment* and say, "Aha, I bet this is important because I know we are about to do an experiment." Finally, you might place a check mark after the phrase *fair test* and say, "I knew that. I know you have to set up a race so it is fair." Point out that you did not put a code next to every sentence; students only need to place codes where they seem appropriate.

After modeling the strategy for the first two paragraphs, give students time to read and code the rest of the article independently.

Reading Strategy 2 (for "A-Maze-ing Worms"): Group Reading Procedure

To introduce this strategy, hand out What to Do in Your Reading Group to each student. Read over the procedure with your class, then select three confident students to come to the front of the class and model the procedure for the first section of text (down to the first dashed line). Then place the rest of your students into groups of three to try the process themselves. Have them start from the beginning of the article to reinforce the process they just watched.

Remember that the first time students use a procedure like this, it may not go smoothly. Students will benefit most from the process after they have had a chance to practice it two or three times.

> ***TEACHING NOTE***
> I find that giving each group a set of index cards bearing the job titles (interpreter, leader, flag flyer) helps students divide up the roles. Holding a card with a job title seems to help a student remember what role he or she agreed to take.

Journal Questions

What did you do today that helped your reading group? Did you do anything that was unhelpful? What could you do next time to help your group even more?

Application/Post-Reading

- Graphic Organizer: A-Maze-ing Worms: Diagramming Dr. Wheeler's Experiment
- Pulling It Together in Writing—Give students the following prompt: All controlled experiments have some things in common. Write a paragraph comparing your mealworm experiment to Dr. Wheeler's experiment with *C. elegans* worms. How were your experiments similar? How were they different? Think about what you wrote, then answer this question: What things would you expect both of these experiments to have in common with other controlled experiments?
- Pulling-It-Together Focus Point: All controlled experiments are similar in that they create a fair test in which only one variable changes at a time. In addition, they have a control group with which they can compare their experimental group.

References

Edgely, M. 2007. What is *C. elegans*? Caenorhabditis Genetics Center. *www.cbs.umn.edu/CGC/what.html*

Kiontke, K., and W. Sudhaus. 2006. Ecology of *Caenorhabditis* species. In *WormBook*, ed. The *C. elegans* Research Community. *www.wormbook.org/chapters/www_ecolCaenorhabditis/ecolCaenorhabditis.html*

National Resource Council (NRC). 1996. *National science education standards*. Washington, DC: National Academies Press.

Quin, J., and A. R. Wheeler. 2007. Maze exploration and learning in *C. elegans*. *Lab on a Chip* 7: 186–192.

REMEMBER YOUR CODES

! This is important.
✓ I knew that.
X This is different from what I thought.
? I don't understand.

On Your Mark!

Source: istockphoto

Elijah and Jamie had been arguing since first period about who was the faster runner. After lunch, they finally got a chance to go outside for a race. They decided to run from the back wall of the gym to the nearest tree.

Jamie and Elijah essentially created a simple controlled experiment. They wanted to answer the question "Who is faster?" To get an accurate answer, they first had to make sure that they set up a fair test.

They looked over the course. From where they stood, Jamie would have to run around a bush to get to the finish line, but Elijah would not. "Unfair!" Jamie said. "We should both have a clear path."

They moved over so that neither one had to circle the bush. Jamie edged forward so he was standing a little bit ahead of Elijah. "Now that's not fair to me," said Elijah. "We should both start with one foot against the wall so neither of us can get a head start."

Scientists think about the fairness of their experiments, too. Suppose a group of scientists was testing how well a new medicine controlled acne. They might get two groups of volunteers. One group would get lotion containing the new medicine. This group would be the experimental group. The other group would get a lotion with no medicine in it. This group would be the control group. The data from the control group shows the scientists what would happen normally, without the medicine.

But what if the people in the experimental group also washed their faces with acne soap? It would be hard to know if their acne improved because of the medicine or because of the soap they used. Or what if the control group started out with less acne? It would be hard to compare the improvement in the two groups.

Anything that *could* be different between the groups—such as the medicine they get, the soap they use, or how much acne they have—is called a *variable*. To make the test fair, scientists design their experiment so that only one variable changes between the two groups—the variable they are testing. In the acne experiment, the scientists would want the only difference between the groups to be whether or not the participants took the medicine. The scientists might control the other variables by starting with people who have similar acne and having everyone in both groups use the same soap and washing routine.

Scientists also have to consider how to judge the final results of their experiment. When Elijah and Jamie raced, they agreed that the first person to the tree would be considered faster. For the acne experiment, the scientists might count the number of pimples in a three-week period, or they might count the number of days that the participants were acne-free. At the end of the experiment, they can compare the results for the two groups to determine which group had less acne.

On your mark, get set, go! Elijah and Jamie headed for the tree. Elijah tagged it just a moment before Jamie.

"All right, all right," said Jamie. He knew the race had been a fair test. "You're faster," he said, "for today."

Source: istockphoto

REMEMBER YOUR CODES

! This is important.

✓ I knew that.

X This is different from what I thought.

? I don't understand.

A-Maze-ing Worms

Dr. Aaron Wheeler loaded a small wire loop with bacteria, lowered it into a dish, and waited. In moments, a tiny worm climbed into the loop to eat the bacteria. Dr. Wheeler placed the worm into a maze and started his stopwatch. The worm was tiny—only about the size of a dash (-). But Dr. Wheeler was using the worm to find out more about how people learn.

Why Worms?

When scientists want to learn something about humans, they often study smaller animals first. In this case, Dr. Wheeler's worms are very simple. Scientists know exactly how many cells are in each worm and what each cell does. They know all about the worm's DNA. They even know what chemicals are in the worm's brain. This information makes it easier to figure out cause and effect. If scientists can figure out which chemicals affect learning in the worms, they can see if the chemicals work the same way in humans.

____

The Worms

The scientific name for Dr. Wheeler's worms is *Caenorhabditis elegans,* but scientists call them *C. elegans* for short. In nature, *C. elegans* live in gardens and compost bins. They hunt through the dirt for bacteria to eat and hitch a ride on a millipede or a slug when they want to get to new hunting grounds. They can't hitch a ride in Dr. Wheeler's lab,

A Photograph of *C. Elegans* Taken Through a Microscope

Source: Zeynep F. Altun, editor of *www.wormatlas.org. http://commons.wikimedia.org/wiki/File:Adult_Caenorhabditis_elegans.jpg*

though. They have to find their way through the maze all by themselves.

Pint-Size Puzzles

First, Dr. Wheeler had to build a maze for his tiny worms. He used a machine that normally builds computer microchips to create thin, rubbery mazes about the size of a postage stamp. Each maze was shaped like a *T*, so the worms only had to make one decision: turn right or turn left. But the worms don't have eyes, so even that simple maze was hard for them to figure out.

Dr. Wheeler always placed the worms' food, a clump of bacteria, on the right side of the maze. Each worm had five chances to find the food.

"The first time, the worms averaged about 15 minutes to get to the food," said Dr. Wheeler, "but they finished in about 5 minutes by the fifth try."

Even when Dr. Wheeler stopped putting the food in the maze, the worms kept going back to the same spot. They had learned.

____

Topic: Worms
Go to: *www.scilinks.org*
Code: LSB002

Human Learning

People can learn in much the same way. The first time you read a difficult word, it might take you several tries to figure out what it says. But each time you see the word again, you will figure it out more quickly. Soon you will have learned to read the word instantly.

Some people have a much harder time learning to read. Math, memorizing lists, or figuring out puzzles may come easily for them. But even when they read the same word over and over again, they may have a hard time remembering what it says. This is called a learning disability.

No one knows for sure what causes learning disabilities, but scientists suspect that certain chemicals in the brain may be involved. One major brain chemical is called *dopamine* (pronounced *DOPE-uh-mean*).

"There's a hypothesis," said Dr. Wheeler, "that dopamine helps with learning."

A Chemical Connection

Did dopamine help Dr. Wheeler's worms learn? He decided to find out. Other scientists had a group of worms that were just like the ones Dr. Wheeler was using, except for one thing: The new worms didn't make dopamine.

C. Elegans in a T-Shaped Maze

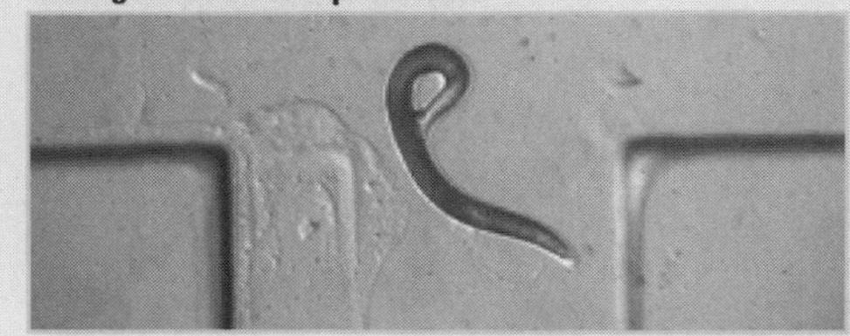

Source: Aaron Wheeler.

Dr. Wheeler repeated his experiments. He used normal *C. elegans* as his control group. He measured how fast they found the food and how long they remembered where to go after the food was gone. Then he tested the worms without dopamine. Those worms were able to find the food fairly well. But when the food was removed, the worms without dopamine forgot their knowledge much more quickly than the normal worms.

"This research supports the hypothesis that dopamine helps with learning," said Dr. Wheeler.

"In the future, we may be able to use that knowledge and"—he gestured to the worm in the maze—"these little guys to help people with learning disabilities."

THE BIG QUESTION

What question did Dr. Wheeler want to answer in his experiment? Based on the information in this article, how would you answer Dr. Wheeler's question?

A Wormy Question: Design Your Own Controlled Experiment

Background Information
What do you already know about mealworms?

Question
How do mealworms respond to _________?

Methods
What will you do? (answer in the sections below)

Groups
What will your experimental group(s) be?

What will your control (normal) group be?

Procedure
Draw and label a diagram of how you will set up your experiment.

How many times will you do your tests?_____

What things will you do to make sure you have a fair test?

Specifically, what will you measure to compare the experimental and control groups? (You may want to use words like *how many*, *how far*, *how much*, or *how long*.)

Prediction
What do you think will happen? Why do you think that?

Results
Make a table to fill in as you conduct your experiment.

Conclusions
Make a Claim: What is your answer to the question you posed in Question 2?

What evidence can you give to support your answer?

What to Do in Your Reading Group

1. The *interpreter* finds "The Big Question" at the end of the reading selection and reads it aloud. Remember, this is what you are trying to learn!

2. Everyone reads the first section quietly and marks ✓, ?, !, and x while reading.

3. The *leader* asks each member of the group to share anything that was confusing (marked ? or x).

4. The group should try to figure out what the confusing word, sentence, idea means. If the group cannot make sense of the confusing word, sentence, or idea, the *flag flyer* should raise the flag.

5. Repeat steps 1–4 for Section 2.

6. Repeat steps 1–4 for Section 3.

7. The group should work together to answer The Big Question. The *interpreter* will write the group's answer to turn in to the teacher.

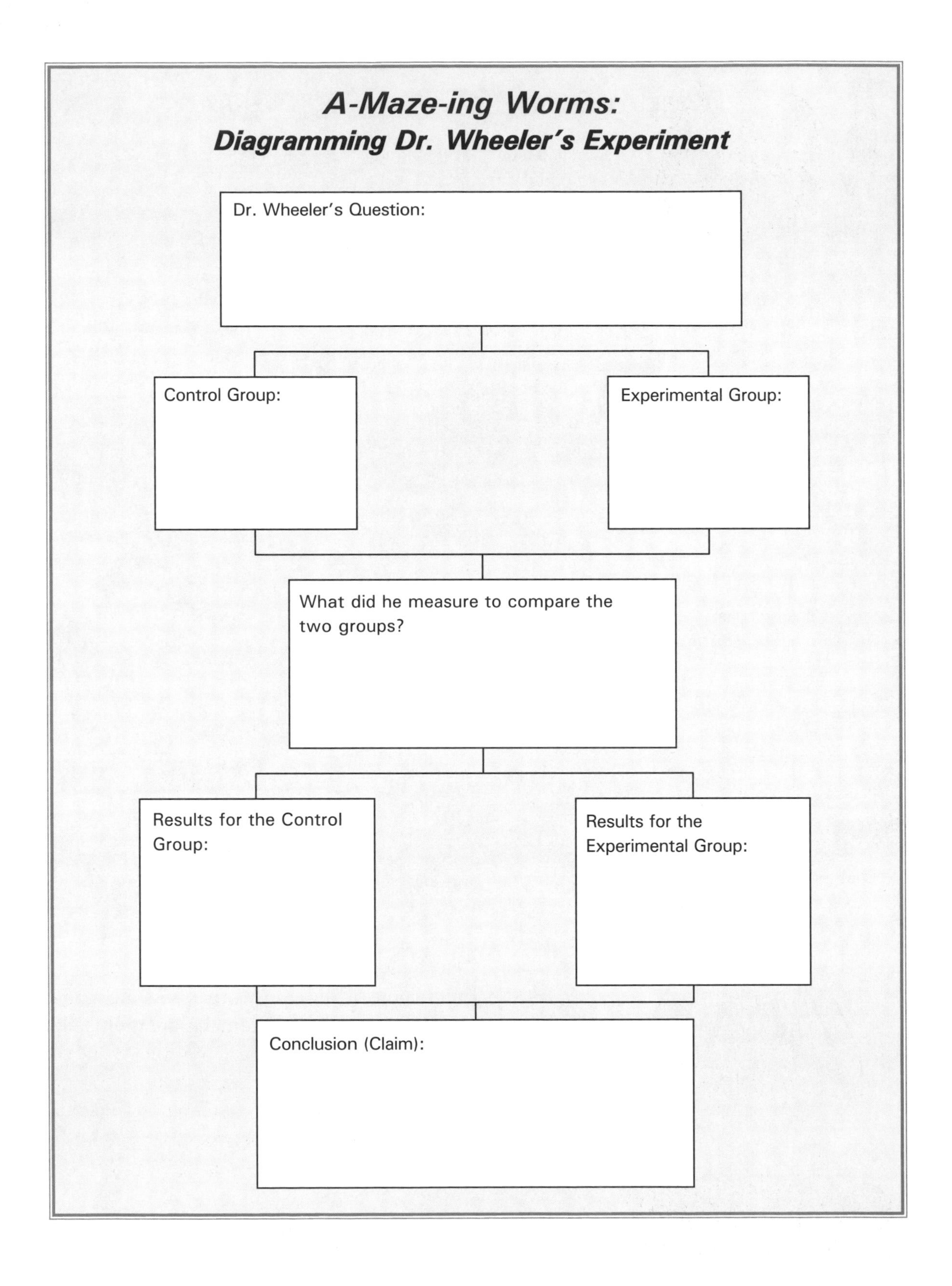
A-Maze-ing Worms:
Diagramming Dr. Wheeler's Experiment
Dr. Wheeler's Question:
Control Group:
Experimental Group:
What did he measure to compare the two groups?
Results for the Control Group:
Results for the Experimental Group:
Conclusion (Claim):

Chapter 5

Cells R Us

Topics

- Plant, animal, and bacteria cells
- Cell parts
- Prokaryotes and eukaryotes

NSES Content Standards (For Grades 5–8, Life Science)

- Living systems at all levels of organization demonstrate the complementary nature of structure and function. Important levels of organization for structure and function include cells, organs, tissues, organ systems, whole organisms, and ecosystems.
- All organisms are composed of cells—the fundamental unit of life. Most organisms are single cells; other organisms, including humans, are multicellular.
- Cells carry on the many functions needed to sustain life. They grow and divide, thereby producing more cells. This requires that they take in nutrients, which they use to provide energy for the work that cells do and to make the materials that a cell or an organism needs. (NRC 1996, p. 156)

Reading Strategy

Using context clues to find the meaning of new words

Topic: How Do Plant and Animal Cells Differ?
Go to: *www.scilinks.org*
Code: LSB003

Topic: Bacteria
Go to: *www.scilinks.org*
Code: LSB004

Background

This lesson is designed to open a unit on cells and cell parts. In the exploration, students will view plant, animal, and bacteria cells through a microscope while looking for the answers to their own questions about cells. Afterward, the reading will provide essential cell-related vocabulary that the students can apply to the visual images they created for themselves during the exploration.

Materials

- Clean slides and coverslips
- Droppers
- Toothpicks
- Methylene blue stain (available from biological supply companies)
- Live elodea plants (available in the aquarium section of many pet stores as well as from biological supply companies)
- Prepared slides of *E. coli* or other bacteria and cheek cells
- Microscopes
- Indirectly vented chemical splash goggles, vinyl gloves

Student Pages

- "Cells R Us"
- Cells: The Inside Story
- At Work in a Cell

Exploration/Pre-Reading

In this exploration, students will examine and draw human cheek cells, cells from the leaf of the aquatic plant *Elodea*, and cells of *E. coli* or other bacteria.

The cheek cell slides and *Elodea* slides should be prepared in advance by the teacher for safety. Each group will need one prepared cheek cell slide and one *Elodea* slide. You may need to add water to keep them fresh throughout the day.

The plant and animal cells can be seen most clearly at 100× to 400×. The bacteria cells can be seen best through an oil immersion microscope.

SAFETY ALERT!

- When working with methylene blue stain, wear goggles and gloves. Review Material Safety Data Sheet (MSDS).
- Wash hands with soap and water upon completion of activity.

If you have access to an oil immersion microscope, you can set up one for the class and allow all students to rotate through to view it. Otherwise, simply use the bacteria slides at the highest setting of the microscopes you have available.

To make a cheek cell slide for your students, place a drop of water on the slide. Gently scrape the inside of your cheek with a toothpick and rub the toothpick in the drop of water. Lower the coverslip, being careful not to trap air bubbles in the water. Then place a tiny drop of methylene blue against the edge of the coverslip. Hold a bit of paper towel against the coverslip on the opposite side to absorb water. This will draw the stain across the slide and provide an even color. Make at least one slide that is not stained so that students can see the original color of the cell.

To make an *Elodea* slide, select a young leaf from near the top of the plant. These young leaves are only a few cells thick. Place the leaf in a drop of water on the slide and add a coverslip. Once again, be careful not to trap air bubbles, as students tend to mistake these for cells.

Begin class by asking students what they know about cells and compile a list on the board. (If you have students who are learning English, you may need to explain that the scientific word *cell* is different from a prison cell or a cell phone.) Tell students about the three cells they will view. Ask them to think about questions they might be able to answer by looking at those cells and list the questions on the board. Possible questions include

- What shape are cells?
- What color are cells?
- Do cells move?
- What's inside a cell?
- How are plant, animal, and bacteria cells different from each other?

Hand out Cells: The Inside Story. Have students record three things they already knew about cells or learned from the class list. Then place students in their lab groups. Each group should select a specific question that it would like to answer during the lab. Note that all students will perform the same procedures; the question the group chooses will determine how students use the information they gather in the lab. It is fine for multiple groups to investigate the same question.

Next, explain the procedures for the lab. If the slides are premade, explain how they were made. Otherwise, demonstrate for students how to prepare the slides. If students are not familiar with procedures for making careful field notes and drawings, stress that the cells should be drawn large enough to fill most of the available space on the lab sheet so they have room to include details. If they know the names of any of the parts they see, they can label them, but it is perfectly acceptable if they do not know any names.

Topic: Prokaryotic Cells
Go to: *www.scilinks.org*
Code: LSB005

When students have completed the lab, they should attempt to answer the question they selected at the beginning of the lab, using their lab notes. These answers do not need to be sophisticated. For example, if a group asked, "What shape are cells?" an acceptable answer would be that they seem to be different shapes. For evidence, they could say that the plant cell was a rectangle, the bacteria was an oval (rod shape), and the cheek cell was a blob.

Introduce the Reading. Explain that reading is another way that scientists investigate things. The next step in students' cell investigation is to learn more about what they saw under the microscope by reading about it.

Reading Strategy: Finding the Meaning of New Words

To introduce this strategy, ask students if they have ever come across a new word in a science book. Tell them that it can be hard to read science books because of all of the new words, but the nice thing about science writing is that the text usually tells you what the new words mean. You just have to know how to recognize the definitions.

Display Table 2.1 from page 14 so that students can see it. Have students read through the table, then ask students if they can think of other ways a definition might be given, adding any of their ideas to the chart. Point out that definitions are usually given just before or after a new word is used for the first time. Have students watch for new words and their definitions as they read.

Journal Question

In today's reading, you looked for ways that definitions can be given for new words. Now you try writing a paragraph that includes a definition. Think of a slang word that an adult might not know (that is *not* offensive). Using one of the examples on the board as a pattern, write a paragraph that introduces the meaning of your word.

Application/Post-Reading

- Graphic Organizer: At Work in a Cell
- Pulling It Together in Writing: Give students the following prompt: Agent Mildew has just started working for the FBI. It's his first day on the job, and he has been given some cells that were collected from a crime scene. He needs your help to figure out if the cells come from a plant, animal, or bacteria. Write a letter to Agent Mildew explaining what he should look for under the microscope to know what kind of cells he has.
- Pulling-It-Together Focus Point: The presence of a large vacuole, chloroplasts, and a cell wall indicates a plant cell. The absence of these cell parts, in a cell with a nucleus, indicates an animal cell. A small cell without a nucleus would be bacterial.

Reference

National Resource Council (NRC). 1996. *National science education standards*. Washington, DC: National Academies Press.

REMEMBER YOUR CODES
- ! This is important.
- ✓ I knew that.
- X This is different from what I thought.
- ? I don't understand.

Cells R Us

Topic: Cell Structure
Go to: *www.scilinks.org*
Code: LSB006

If you were to peer through a microscope at a thin layer of your skin, you would see that skin is not just a solid sheet that covers your body. Skin is divided into a network of tiny cells. And it's not just your skin—every part of your body is made up of cells. Each cell is like a tiny living thing by itself, and all of your cells work together to keep your body going.

Cell Structure

The first part of a cell that you might notice under a microscope is the cell membrane. This protective covering has tiny openings, called *pores*, that let things enter and leave the cell. Some things, such as water, move in and out of the cell easily. Other things, such as most proteins, are only allowed in and out when needed.

If you were a plant, each of your cell membranes would be surrounded by a stiff outer layer, called the *cell wall*. Cell walls are tough and flexible and help plant cells hold their shape. The cell wall helps give celery its crunch and dead leaves the ability to crumble in your hands. Animal cells don't have cell walls. They rely on the cell membrane for shape and support.

If you look at your skin through a light microscope, it might look like your cells are mostly empty, but nothing could be further from the truth. Powerful electron microscopes reveal that cells are packed with

Parts of a Typical Plant Cell

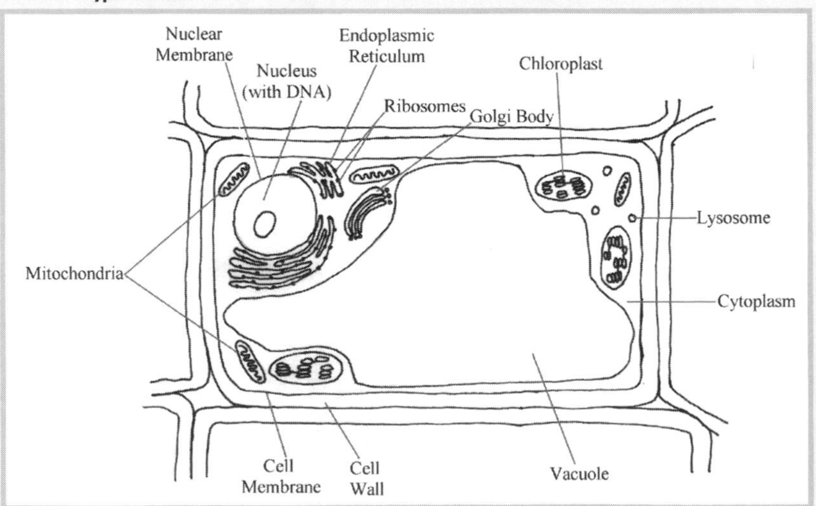

tiny parts that are constantly in action. These parts are called *organelles* because they function like organs for your cells. Just like your heart has the job of pumping blood and your intestines have the job of digesting food, each organelle in the cell has a specific job to do. And just like your organs are surrounded by layers of fat and liquid, the organelles in your cells are cushioned by a watery gel called cytoplasm.

Manufacturing and Transport Organelles

One of the main functions of a cell is to make proteins. This job falls to the *ribosomes*, which are the cell's manufacturing plants. After the proteins are made, they are shipped to wherever they are needed through a network of tubes called the *endoplasmic reticulum*, or ER. If the proteins need to be shipped outside of the cell, they are packaged up for transport in another organelle, the Golgi body.

Energy Organelles

All of this manufacturing takes energy. *Mitochondria* are the organelles that produce energy. Mitochondria work like an oil refinery. Oil has a lot of energy, but it has to be made into gasoline before it can power a car. Mitochondria take the energy from the food you eat and convert it into a power source that the other organelles can use.

Many plant cells can even get energy right from the source. An organelle in plant cells, called the *chloroplast*, captures energy from sunlight to use to make sugar. The plant doesn't need to eat food to fuel its cells. It simply ships the sugar over to the mitochondria for processing.

Waste and Storage Organelles

Two other organelles, *vacuoles* and *lysosomes*, help with storage and waste removal. Vacuoles serve as storage tanks, holding water or nutrients until they are needed or storing waste until the cell is ready to get rid of it. Lysosomes actually digest waste. They break down old cell parts and invading bacteria and viruses. The cell can sometimes reuse pieces of this waste after it has been digested.

SCILINKS THE WORLD'S A CLICK AWAY
Topic: DNA
Go to: *www.scilinks.org*
Code: LSB007

Parts of a Typical Animal Cell

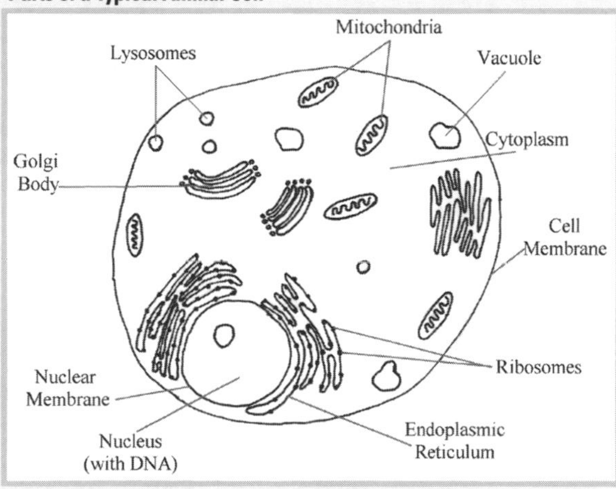

Parts of a Typical Bacterial Cell

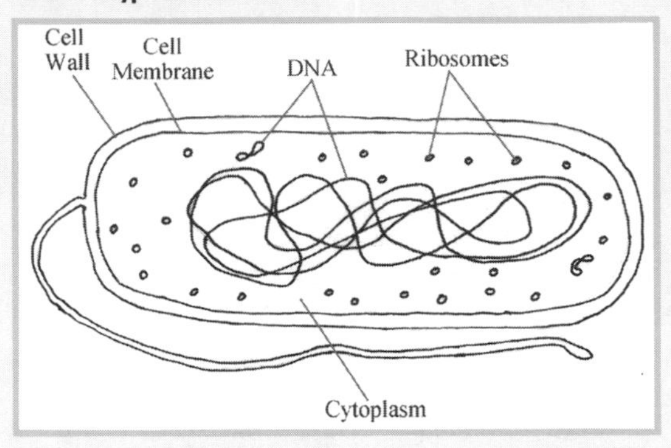

Organelle in Charge

To work together smoothly, the organelles need a director. This is the job of the *nucleus*. The nucleus contains the chromosomes, which are made of deoxyribonucleic acid, or DNA. You've probably heard of DNA. It gives the instructions for all of the activities of the cell, including the recipes for making different types of proteins. The nucleus is surrounded by its own membrane, the nuclear membrane, which holds all of the chromosomes in place.

Those Weird Bacteria

The organelles discussed above are common in plant and animal cells and cells from fungi, such as mushrooms. Bacteria cells, however, are different. They are much smaller than other cells. For example, about 30 of the bacteria *E. coli* would fit inside one of your skin cells.

Bacteria have cell membranes, and some have cell walls. They use ribosomes to make protein. But bacteria cells don't have any organelles with their own membranes, like most of the organelles previously discussed. They don't even have a nucleus. The chromosomes in bacteria float loosely through the cell.

Organisms whose cells have a nucleus, such as plants and animals, are called *eukaryotes*, a word that means true nucleus. Bacteria are called *prokaryotes*, or before nucleus, because bacteria cells developed before eukaryotic cells.

Cells All Over

Those skin cells you see under the microscope are just a sampling of the trillions of cells that make up your body. Those cells have a lot in common with cells from organisms around the world. From the tiniest bacteria to the biggest animals and plants, all living things depend on the activities of their cells to survive.

THE BIG QUESTION

Imagine you were looking at cells from a pine needle through a powerful electron microscope. Based on the reading, what cell parts would you expect to find?

Cells: The Inside Story

Part 1: Background Knowledge

What are some things that you know about cells?

Part 2: Question

Write the question that you select here:

Why did you pick this question?

What do you think the answer is?

Part 3: Procedures

You will look at three typical cells under the microscope. Draw one cell from each slide. Make your drawings BIG to show details, and note colors, movement, and other information that might help you answer your question. If you know the name of a cell part, label it. Be sure to write how many times the cell has been magnified.

Part 4: Data

Animal Cell: Look at the prepared slides of cheek cells under the microscope. They have been dyed blue so that more parts can be seen. The cheek cell has been magnified _________ times its normal size.

Draw the cheek cell and label it as needed.	Animal cell notes

Look at the cheek cell that has not been stained. What color is it? ________________

Plant cell: Look at the leaf from the *Elodea* plant under the microscope. They have not been dyed. Each rectangle you see is one cell. Draw one cell below. The *Elodea* cells have been magnified __________ times their normal size.

Draw the *Elodea* cell and label it as needed.	Plant cell notes

Bacteria Cell: Look at the prepared slides of bacteria. These bacteria are dead and have been dyed. You will not be able to see as much detail in these cells because they are much smaller. These cells have been magnified _______________ times their normal size.

Draw a line of bacteria cells and label it as needed.	Bacteria cell notes

Part 5: Conclusions

A. Claim: Look at the question you wrote for Part 2. Based on what you have observed, try to answer that question.

B. Evidence: What evidence supports the claim you made above? That is, what data from the lab supports the answer you gave?

At Work in a Cell

Category	Name of the Cell Part	Function
Cell Structure		
Manufacture and Transport		
Energy Production		
Waste and Storage		
Directing		

Chapter 6

Healing Powers

Topics

- Cell cycle
- Mitosis

NSES Content Standard (For Grades 5–8, Life Science)

- Cells carry on the many functions needed to sustain life. They grow and divide, thereby producing more cells. This requires that they take in nutrients, which they use to provide energy for the work that cells do and to make the materials that a cell or an organism needs. (NRC 1996, p. 156)

Reading Strategy

Previewing diagrams and illustrations

Background

Topic: Mitosis
Go to: *www.scilinks.org*
Code: LSB008

Mitosis and cell division are the remarkable means by which the body grows and repairs damaged tissue. The basic ideas in this chapter are simple: The cell duplicates important parts (including DNA) and then divides in half. However, students often find the topic of cell division to be difficult because they become bogged down in the immense vocabulary associated with it. This lesson begins with a video to help students develop a mental image of the process, then the article and activities help them master the key words. In some ways, this resembles how scientists first learned about mitosis. Scientists observed the phenomenon first, watching it under the microscope, then gave names to what they saw.

Material

- A short video animation of mitotic cell division (see exploration section below)

Student Pages

- "Healing Powers"
- The Cell Cycle

Exploration/Pre-Reading

In this exploration, students will watch a video animation of mitosis and cell division that is played without sound. As they watch the video, they will describe what they see in their own words.

Before class, find a video animation of mitosis that has minimal written descriptions or vocabulary on the screen. The narration is unimportant, as you will turn the volume off when students watch it. Several such videos are available online, including

- *www.johnkyrk.com/mitosis.html* (This is a particularly clear and simple video. Instruct students to ignore the phase names posted to the left of the animation.)
- *www.cellsalive.com/mitosis.htm* (Have students focus on the animation in the middle, not the surrounding text and information.)

- *www.youtube.com/watch?v=VlN7K1-9QB0* (Many schools block YouTube, but this is a very detailed video if you can access it.)

You may also have a video in your school or classroom library that includes a segment that would work well.

Start class by telling students that you are going to show them a video with no sound. They should describe in writing whatever they see happening. If they think they know some of the science words related to what they see, they may use them, but they do not need to. Explain that you will show the video several times so that they can complete their description. Make sure the volume is turned off, and direct students to ignore any words that may show up on the screen.

Show the video clip three or four times to ensure that students are able to write a complete description and form a mental image of the process of cell division. You may then ask for volunteers to read their narration aloud as you show the video one or two more times.

Introduce the Reading. Ask students to guess what process the video was showing. If someone guesses that it was a cell dividing in half, confirm their answer. Otherwise, tell the class that they were watching how a cell divides. Explain that the process of cell division is fairly simple, but scientists use a lot of vocabulary words to describe it. The text they are about to read will explain the process and teach them the words that scientists use when talking about cell division.

> ***TEACHING NOTE***
> "Healing Powers" includes the most common vocabulary words associated with mitosis. However, some state curriculum guidelines allow teachers to reduce the vocabulary demands of this topic by leaving out the names of the four phases. If that is the case where you live, encourage students to focus on understanding the main ideas without worrying about those particular words.

Reading Strategy: Previewing Diagrams and Illustrations

Tell students that in some books that they read, the pictures may be extras. In science writing, however, the pictures and diagrams often carry a lot of important information. Looking at the pictures and making predictions about what they mean before reading can help make the text easier to understand.

Hand out "Healing Powers," but tell students not to read it yet. Look through the figures as a class, providing the following prompts to help them study the diagrams and predict their meanings. For the prediction questions, accept any answer as a valid possibility, and tell students that they will have to read the text to find out if they are right.

Start with Figure 1, page 57. Ask:

- Does this look like anything you saw in the video?
- What does the diagram tell you about the *X*s you saw in the video? What are they called? (chromosomes)
- What do you think the diagram is showing in Part B? (prediction)

Next, direct them to look at Figure 2 (p. 57).

- What new cell parts are shown in that diagram? (centrioles and spindle fibers)
- What might those parts do? (prediction)

Then have them look over Figures 3 through 5 (p. 58).

- Do you notice anything that seems different from what you saw in the animation? (This question allows you to comment on different ways things were represented, in case something about the representation in the text is confusing.)
- What cell part reappears in Figure 5? (nuclear membrane)

Finally, have them compare Figure 2 with Figures 5 and 6 (p. 58).

- How many cells were there to start? (1)
- How many cells were there in Figure 6? (2)
- How many chromosomes were in the cell originally? (4)
- How many are in each nucleus at the end? (4)

Journal Questions

Do you typically study the diagrams and pictures in your science book before you read? Was it helpful to look at the diagrams and pictures before reading today? Why or why not?

Application/Post-Reading

- Graphic Organizer: The Cell Cycle
- Pulling It Together in Writing: Have students take out the descriptions they wrote of the video before reading. They should now know most of the correct vocabulary to describe cell division. Have them rewrite their description to include their new vocabulary. You can indicate that they are "translating" from everyday language to science language. You may wish to show the video again and allow volunteers to read their new narration.
- Pulling-It-Together Focus Point: Chromosomes in the cell double and move to opposite sides of the cell. Then the cell divides, and each new cell has the same number of chromosomes as the original.

Reference

National Resource Council (NRC). 1996. *National science education standards*. Washington, DC: National Academies Press.

REMEMBER YOUR CODES

! This is important.

✓ I knew that.

X This is different from what I thought.

? I don't understand.

Healing Powers

"Watch this!" called Andre as he skated past Marie. He raced up the ramp and went into a spin. He felt himself leaning too far sideways, but it was too late to right himself. Crash! He hit the concrete and slid.

Marie hurried over to help him up. "That's gotta hurt," she said, looking over a long scrape on Andre's leg.

Andre took off his helmet and stared at his leg. He needed some new skin.

Where Does New Skin Come From?

Skin is made of cells, like all of the tissues in the human body. Fortunately for Andre, skin cells are constantly being brushed off and need to be replaced. So the repair process was already under way. To make new skin, a surviving cell divides in half to create two identical cells.

Getting Ready to Divide

Even before the accident, Andre's cells were getting ready to divide. Outside the nucleus, the cells got larger and grew more organelles. Inside the nucleus, they made extra copies of their DNA. Each chromosome replicated, resulting in two identical copies that stayed stuck together in one spot. As shown in Figure 1, these copies are called *chromatids*, and the place where they are stuck together is called the *centromere*.

Dividing the Nucleus

The nucleus divides first, in a process called *mitosis*. Scientists separate mitosis into four parts: *prophase*, *metaphase*, *anaphase*, and *telophase*.

Figure 1.

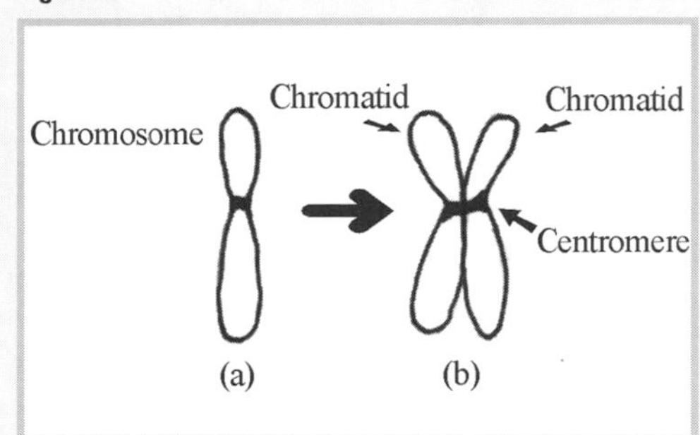

In prophase, the chromatids shorten and thicken so much that they can be seen under a microscope. The nuclear membrane breaks down so there is plenty of room for the chromatids to move around. A pair of organelles called *centrioles* head for opposite ends of the cell. These organelles will work a bit like fishing rods. They create a bridge of spindle fibers between them that will later be used to pull the chromatid pairs apart.

Figure 2. Prophase

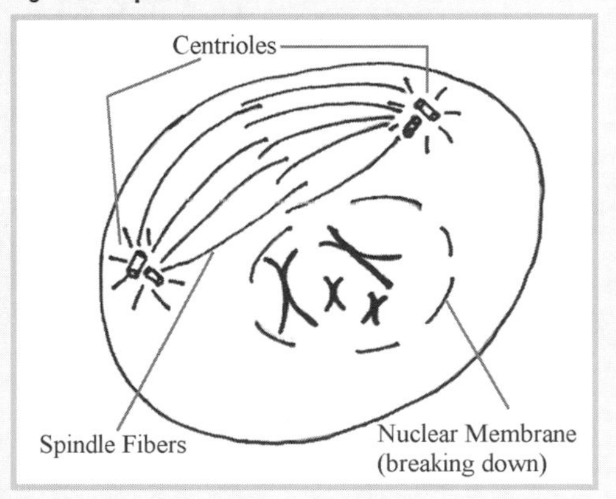

In metaphase, the chromatid pairs line up across the middle of the cell and connect to two spindle fibers at their centromere, the spot where they are joined together.

Figure 3. Metaphase

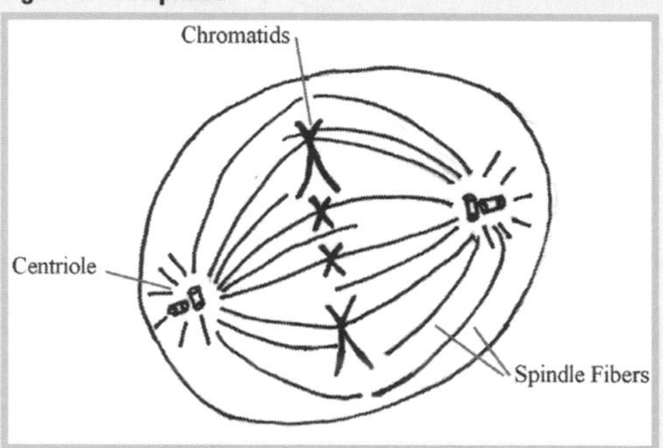

Next comes anaphase, when the centrioles do their work. The centrioles use the spindle fibers to separate the chromatid pairs and pull them to opposite sides of the cell. At the end of anaphase, there is one complete set of chromosomes at each end of the cell.

Topic: Cell Cycle
Go to: *www.scilinks.org*
Code: LSB009

Figure 4. Anaphase

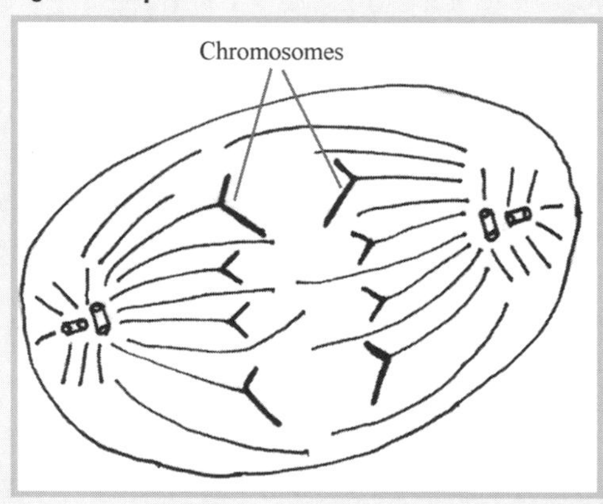

Finally, in telophase, a nuclear membrane forms around each set of chromosomes. The chromosomes lengthen and get thinner until they can no longer be seen through a microscope.

Figure 5. Telophase

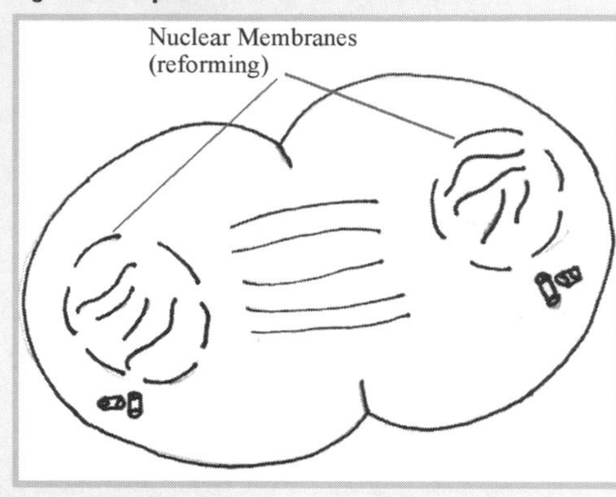

Dividing the Cell

At the end of mitosis, the cell is extra large and has two identical nuclei. All that is left is for the cell to divide in half. The process of cell division is called *cytokinesis*. After cytokinesis, the original cell has become two daughter cells with exactly the same DNA.

Figure 6. Cytokinesis

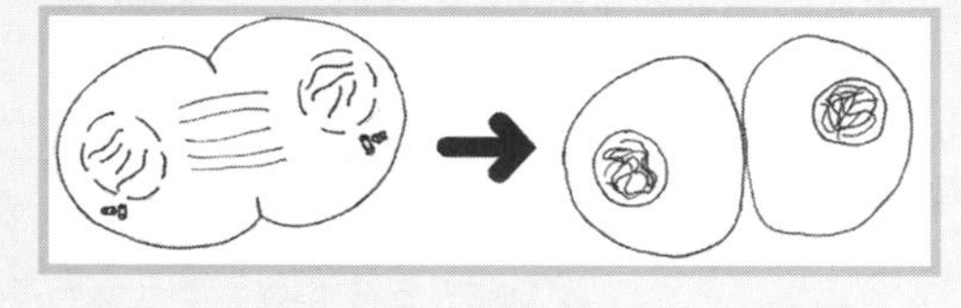

The Cell Cycle

Because of Andre's fall, he needed a bunch of new cells at once. But even without an injury, cells are always in the process of growing, getting ready to divide, and then dividing. This life cycle of a cell is called the *cell cycle*. Some cells only divide occasionally. They spend most of their time in the growing and working phase, called interphase. But tissues that need frequent repair, such as skin and muscle, go through the cycle continually. Each skin cell completes the cycle in about 16 hours. This same cell cycle is what lets a single fertilized egg grow into an adult with trillions of cells.

Healing Powers

Andre sighed as he limped home on his injured leg. "It's going to take forever for this to heal," he thought. But even as he walked, his cells were whizzing along through the cell cycle. His leg would be good as new in just a few days.

THE BIG QUESTION

If Andre asked you how his body gets new cells, what would you tell him?

The Cell Cycle

Cells are constantly moving through the cell cycle. Use the reading selection "Healing Powers" to fill in the phase names where there are blanks.

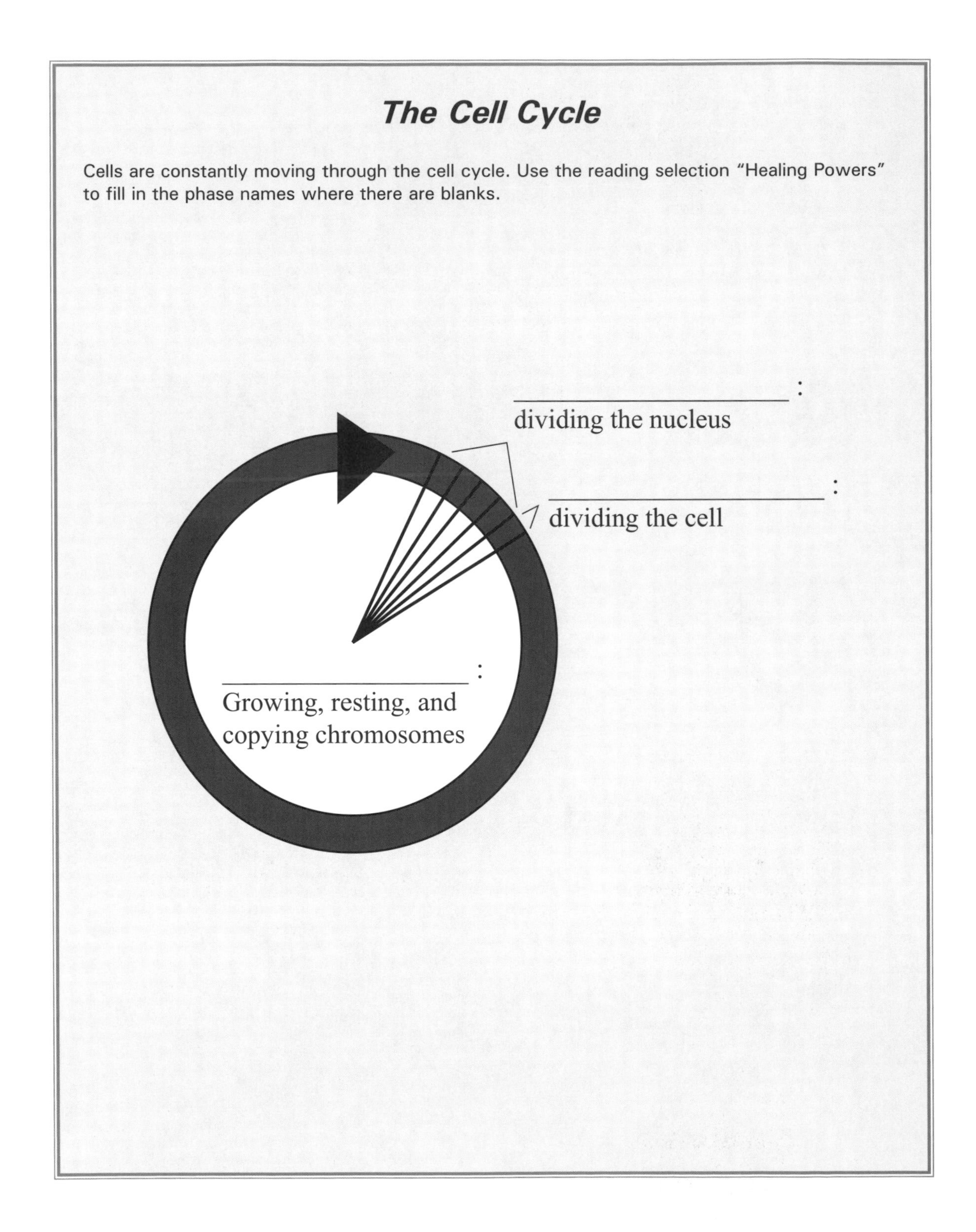

Chapter 7

No Bones About It

Topics

- Characteristics of arthropods
- Invertebrates
- Skeletal and muscular systems

NSES Content Standards (For Grades 5–8, Life Science)

- Living systems at all levels of organization demonstrate the complementary nature of structure and function. Important levels of organization for structure and function include cells, organs, tissues, organ systems, whole organisms, and ecosystems.
- Specialized cells perform specialized functions in multicellular organisms. Groups of specialized cells cooperate to form a tissue, such as a muscle. Different tissues are in turn grouped together to form larger functional units, called organs. Each type of cell, tissue, and organ has a distinct structure and set of functions that serve the organism as a whole.
- Millions of species of animals, plants, and microorganisms are alive today. Although different species might look dissimilar, the unity among organisms becomes apparent from an analysis of internal structures, the similarity of their chemical processes, and the evidence of common ancestry. (NRC 1996, p. 156, p. 158)

Topic: Bones and Muscles
Go to: *www.scilinks.org*
Code: LSB010

Reading Strategy

Identifying text signals for examples and lists

Background

All animals have some form of a skeleton. This skeleton provides shape and support, protects internal organs, and gives the muscles a structure to pull against for movement. Fish, amphibians, reptiles, birds, and mammals have an internal skeleton, made of bones or cartilage. Arthropods have a stiff external skeleton. Worms and other soft-bodied invertebrates have a hydrostatic skeleton in which water pressure provides firm tissues for muscles to pull against. This lesson will compare exoskeletons and internal bony skeletons while also introducing the other main characteristics of arthropods. The lesson should be used to introduce the concept of invertebrates.

SAFETY ALERT!

- Make sure the specimen is pinned securely to the dissecting tray. Do not dissect specimen when held in the hands.
- Point and push sharp dissection tools away from your body.
- Never use excess force when working with a scalpel.
- Use appropriate ventilation to eliminate preservative fumes.
- No food or drink is allowed during dissection activities.
- The teacher is to demonstrate safe procedures for using dissection equipment.
- Wash hands with soap and water after the dissection activity.

Materials

- Large grasshoppers for dissection (1 per student group)
- Dissection scissors, probes, magnifying glasses, and trays
- Indirectly vented chemical splash goggles
- Surgical gloves (vinyl) and aprons
- Model or large diagram of the human skeleton

Student Pages

- "The Arthropods: Characteristics of Phylum Arthropoda"
- Dissecting a Grasshopper!
- All About Arthropods

Exploration/Pre-Reading

In this exploration, students will dissect a grasshopper to observe the major characteristics of arthropods and discover the nature of the grasshopper's skeleton.

Begin by reviewing key aspects of the human skeleton with students. Show the model or diagram of the human skeleton, and discuss the background information questions on the student sheet as a class.

- Find a place where two bones meet. What is that called? (a joint)
- What does having joints allow us to do? (bend or move)
- Look at the backbone. Is it really just one bone? (no)
- What are some functions of the skeleton? (gives shape, allows muscles to attach, etc.)

Topic: Insects
Go to: *www.scilinks.org*
Code: LSB011

Place students in their groups and make sure they have selected roles before giving out the materials. Most of the dissection instructions are basic and simply call attention to key features of the grasshopper. Questions 13 and 14, however, require students to determine what makes up the skeleton of a grasshopper. For Question 13, students will search for bones inside the grasshopper. Some groups may identify a thin spine along the side as a bone. This is actually a ridge that has broken off the inside of the exoskeleton where a muscle connects. Show students that you can press this back against the outer covering where it belongs. Even if they have heard at some point that grasshoppers are invertebrates, they are often surprised when they do not find bones.

SAFETY ALERT!
Make sure that all students wear indirectly vented chemical splash goggles during the dissection. The liquid preservative in the grasshopper can squirt unpredictably.

For the conclusion, allow students to give their best guess as to the nature of the grasshopper's skeleton, supporting it with evidence from their dissection.

Introduce the Reading. Tell students that the text they will read will give them more information on the features they saw on their grasshopper. The reading will also allow them to check their conclusion from the lab.

Reading Strategy: Text Signals—Examples and Lists

Begin by displaying the following excerpt from the reading:

Antennae are appendages that contain sense organs.

Tell students that when you read that sentence, you wondered, "What exactly is a sense organ?" Then you looked at the next sentence. (Write it on the board or uncover it.)

For instance, the antennae of a lobster contain organs for tasting, feeling, and smelling.

Explain that the second sentence gave you an example that helped you understand what a sense organ is: something that lets an animal use its senses, such as tasting, feeling, and smelling.

Tell students that certain words and phrases are signals for what the text is about to explain. Underline the words *for instance*, and explain that *for instance* is a text signal. It tells you that the next few words will give you an example.

Ask students what other text signals they can think of that might indicate that the text is about to give an example. They may generate words and phrases such as *for example, e.g., for instance, such as, like, including,* and *to illustrate.*

Next, place this excerpt from the text on the board, and ask students to predict what they think the next few sentences will tell them.

All arthropods share four main features.

Point out that this sentence should signal to the reader that a list is coming. Say that when you read something like this, it tells you to pay attention to the next few sentences so that you will know what the four main features are.

Finally, add the sentence, "Only three poisonous spiders live in the United States: black widows, brown recluses, and hobo spiders." Show students that the colon (:) is another signal that a list is coming.

Before they read, have students scan the article for text signals that indicate examples and lists and underline them. Tell students to pay careful attention to the information given in those places as they read.

Journal Question

Write a sentence or two that tells your favorite foods. Use a text signal to let your reader know that a list is coming. Underline your text signal.

Application/Post-Reading

- Graphic Organizer: All About Arthropods
- Pulling It Together in Writing: Inevitably, some students will be absent on the day of the grasshopper dissection. Allow your class to prepare a summary for those who are out. Have students write a letter comparing and contrasting the human and grasshopper skeletal and muscle systems. They can use the information they learned in lab and from the reading. Give the copies of the best letters to students who were absent.
- Pulling-It-Together Focus Point: Both grasshoppers and humans have skeletal systems and muscular systems that are jointed and allow them to move. The grasshopper's skeleton is on the outside and is called an *exoskeleton*.

Reference

National Resource Council (NRC). 1996. *National science education standards*. Washington, DC: National Academies Press.

REMEMBER YOUR CODES

! This is important.

✓ I knew that.

X This is different from what I thought.

? I don't understand.

The Arthropods: Characteristics of Phylum Arthropoda

If you were a fly, you could flap your wings more than 100 times in one second. If you were a spider, you would hear with tiny hairs on your legs. If you were a lobster, you would grow a whole new skeleton several times a year.

Flies, spiders, and lobsters are classified as *arthropods*. This phylum, or animal group, also includes all of the insects, crabs, ticks, centipedes, and scorpions. All arthropods share four main features. They have skeletons outside their bodies. Their bodies are divided into segments. They have legs, antenna, and other parts that bend in several places. Finally, they have highly developed organs to sense their environments.

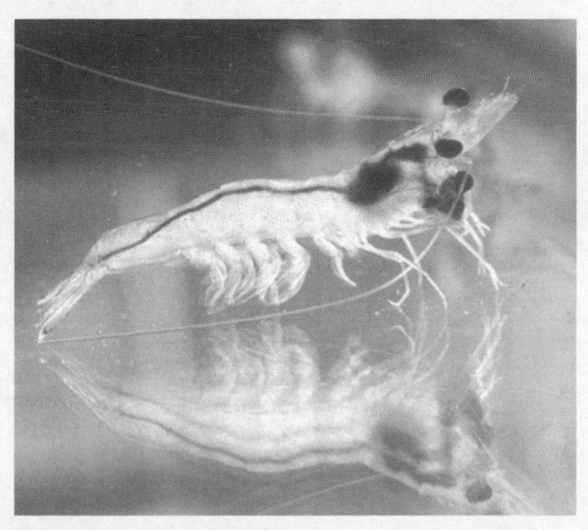

This young shrimp is an arthropod. You can see right through its exoskeleton!

Source: USDA ARS Photo Unit, USDA Agricultural Research Service, *www.Bugwood.org. www.forestryimages.org/browse/detail.cfm?imgnum=1320042*

Arthropod Skeletons

When you think of a skeleton, you probably think of bones. However, not all skeletons are made of bones. An arthropod has a tough, hard covering on the outside of its body. This covering is called an *exoskeleton*, which means skeleton on the outside.

Any skeleton has three main functions: It gives the body shape, protects internal organs, and provides surfaces for muscles to attach. For example, a dog's bones determine its shape. The rib bones protect the dog's heart and lungs. When a dog runs, its leg muscles pull on the leg bones to cause movement. The exoskeleton of a dragonfly serves the same functions. It gives the dragonfly its shape. It protects the soft body inside. The muscles in the dragonfly are anchored to the hard outer covering. Those muscles move the wings and legs to allow the insect to fly and walk.

For their size, exoskeletons are much stronger than bones. This means an insect can fall from a great height without breaking anything. However, an insect's exoskeleton cannot grow once it is formed. This means that insects must replace their outer covering, or molt, as they get bigger.

Animals that don't have bones inside, including arthropods, are called *invertebrates*. Technically, the word *invertebrate* means without a backbone because the bones in the back are called *vertebrae*. However, the word *invertebrate* is used to indicate that an animal has no bones at all.

Segmented Bodies

Arthropod bodies are divided into sections, called segments. In some arthropods, such as centipedes, you can see the segments clearly on the outside. In a centipede, almost every segment has a pair of legs. In other arthropods, the segments are more difficult to see. The segments of an insect are grouped into three regions: the head, thorax, and abdomen. All six legs of an insect connect to the thorax.

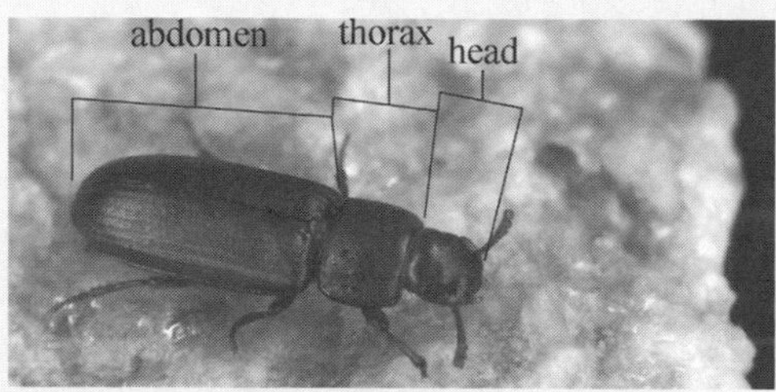

Insects, like this red flour beetle, have three groups of segments.

Source: Peggy Greb, USDA Agricultural Research Service, *www.Bugwood.org. www.forestryimages.org/browse/detail.cfm?imgnum=5174036*

Jointed Attachments

The walking legs of a centipede or insect are just one type of attachment that can connect to the body of an arthropod. Arthropods may also have swimming legs, antennae, claws, or pincers. Many arthropods also have attachments near their mouth to help with slicing or chewing food. All of these attachments are called *appendages*. The appendages in an arthropod have joints that allow them to bend, much like the arms and legs of a person. The name of the phylum reflects this feature. The word *Arthropoda* means jointed foot.

Highly Developed Sensory Organs

Many arthropods have antennae on their heads. *Antennae* are appendages that contain sense organs. For instance, the antennae of a lobster contain organs for tasting, feeling, and smelling.

This ant has appendages on its mouth for cutting leaves.

Source: Susan Ellis, USDA APHIS PPQ, *www.Bugwood.org. www.forestryimages.org/browse/detail.cfm?imgnum=5369293*

Arthropods also have organs for seeing and hearing. The ability to see differs among arthropods. Spiders have eight simple eyes that are primarily used to sense motion. However, many insects have compound eyes that allow them to see color, shape, and movement in all directions at the same time. For hearing, arthropods have organs to detect vibrations. Some arthropods have very sensitive hairs that do this task. Others, such as the grasshopper, have a flexible membrane that works a lot like the ear drum in a human ear.

A Recipe for Success

The arthropods are a diverse group of animals. Their exoskeletons, segmented bodies, jointed appendages, and highly developed sense organs have allowed them to adapt to a wide range of environments. In fact, there are more species of arthropods than of all other animals combined.

THE BIG QUESTION

What features make arthropods unique?

Dissecting a Grasshopper!

Part 1: Background Knowledge

In this lab, you will dissect a large grasshopper, which is in the arthropod phylum. What *kingdom* is the arthropod phylum in? ____________________

You will review some things about the human skeleton before looking at the grasshopper's body.

- Find a place where two bones meet. What is that called? ____________________
- What does having joints allow us to do? ____________________
- Look at the backbone. Is it really just one bone? ____________________
- What are some functions of the skeleton? ________________________________

Part 2: Questions

What makes a grasshopper unique? What makes up the grasshopper's skeleton?

Parts 3 and 4: Data and Procedures

Group Members

A. Operating room director: reads the lab sheet out loud to the group and keeps the group on task.
B. Data transcriber: carefully records the group's findings for others to copy later.
C. Doctor: handles the grasshopper while the group studies the outside of it; holds it steady for the surgeon while the group looks inside it.
D. Surgeon: cuts the grasshopper as needed when the group studies the inside of it.

Grasshopper Body Plan

1. A grasshopper, like all insects, is made up of three main body sections. Use the diagram below to find the three main body sections on your grasshopper.

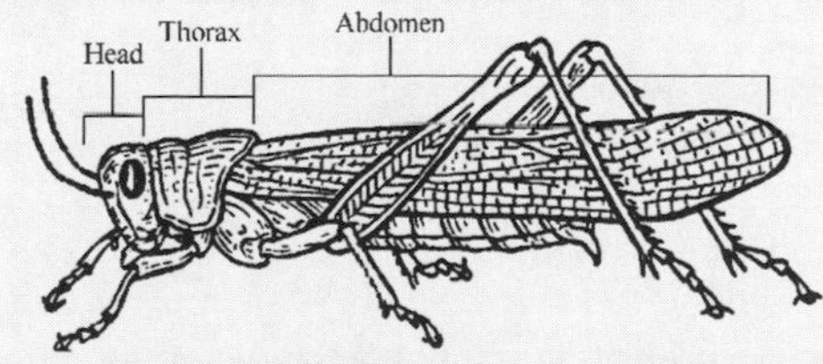

Source: Pearson Scott Foresman. *http://commons.wikimedia.org/wiki/File:Grasshopper_(PSF).png.*

2. All insects have the same number of legs. How many legs does the grasshopper have? _______
   Which body section are the legs connected to? ________________

3. Look at the grasshopper's back legs, which are used for jumping. How many joints are on that leg?_____
   Now, starting at your hip, count how many joints are on a human leg. (Don't forget your toes!)_____

4. The wings on a grasshopper lay folded along the grasshopper's back when it is not flying. Find the wings and gently unfold them. What color is the hind wing? __________
   Which body section are the wings connected to? ____________

Grasshopper Gender

5. You need to figure out if your grasshopper is a male or a female. Female grasshoppers have two pointed body parts sticking off the end of the abdomen. These are used to deposit eggs. The end of the abdomen of male grasshoppers is blunt. Is your grasshopper male or female? __________

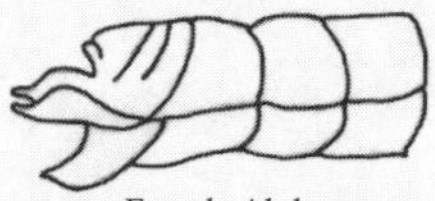

Female Abdomen

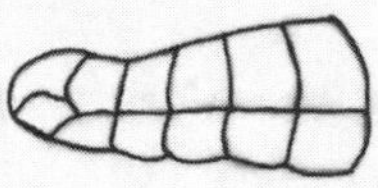

Male Abdomen

Grasshopper Sensory Organs

6. Carefully remove the wings from one side of your grasshopper. Look just above where the grasshopper's jumping leg connects to the body. You should see a circle. This is the *tympanum*, which functions as the grasshopper's ear! It is a membrane that senses vibrations, just like the membrane in our ears that we call the ear drum. What color is the tympanum?________________

Antennae
Simple Eyes
Compound Eye
Chewing Parts

7. Now turn your grasshopper to look at its "face." Grasshoppers have five eyes. Two of them are large, compound eyes. Three of them are called simple eyes because they can only detect light and dark. Find the eyes. You may need a magnifying glass to see the simple eyes.

8. Find the antennae. The antennae are used to feel. Notice that these are also jointed. How would having jointed antennae be more helpful than having straight antennae?
   ____

9. Antennae are also used to sense chemicals. Humans sense chemicals in two places. One of them is on our tongue. What is the other place? (Think what else is on our face ...) ______________________

Inside the Grasshopper

10. Lay the grasshopper on the tray so that the side without wings is facing up. Notice that there is a ridge along the grasshopper's belly. Starting at the tip of the abdomen, cut just to the side of that ridge all the way to the end of the thorax. Try to cut only the outside covering without damaging the organs inside. Do not stick the scissors very deep into the grasshopper.

11. Now cut along the side of the thorax so that you can open the grasshopper.

12. The inside of a grasshopper is mushy and confusing. If you have a female, you may have a covering of eggs over the internal organs. If so, gently pull back the egg layer and cut it off.

The Skeletal and Muscular Systems

13. Gently move the mushy insides around and look for bones. What did you find?
____

14. Gently pull off one of the jumping legs. Cut off the covering and look inside the fattest segment of the leg. You will see a clump of muscle. In humans, muscles connect to bones. What do these muscles connect to? ______________________
____

Part 5: Conclusions

A. Claim: What makes up the grasshopper's skeleton?

B. Evidence: What did you see in your grasshopper that helped you figure out what kind of skeleton it had?

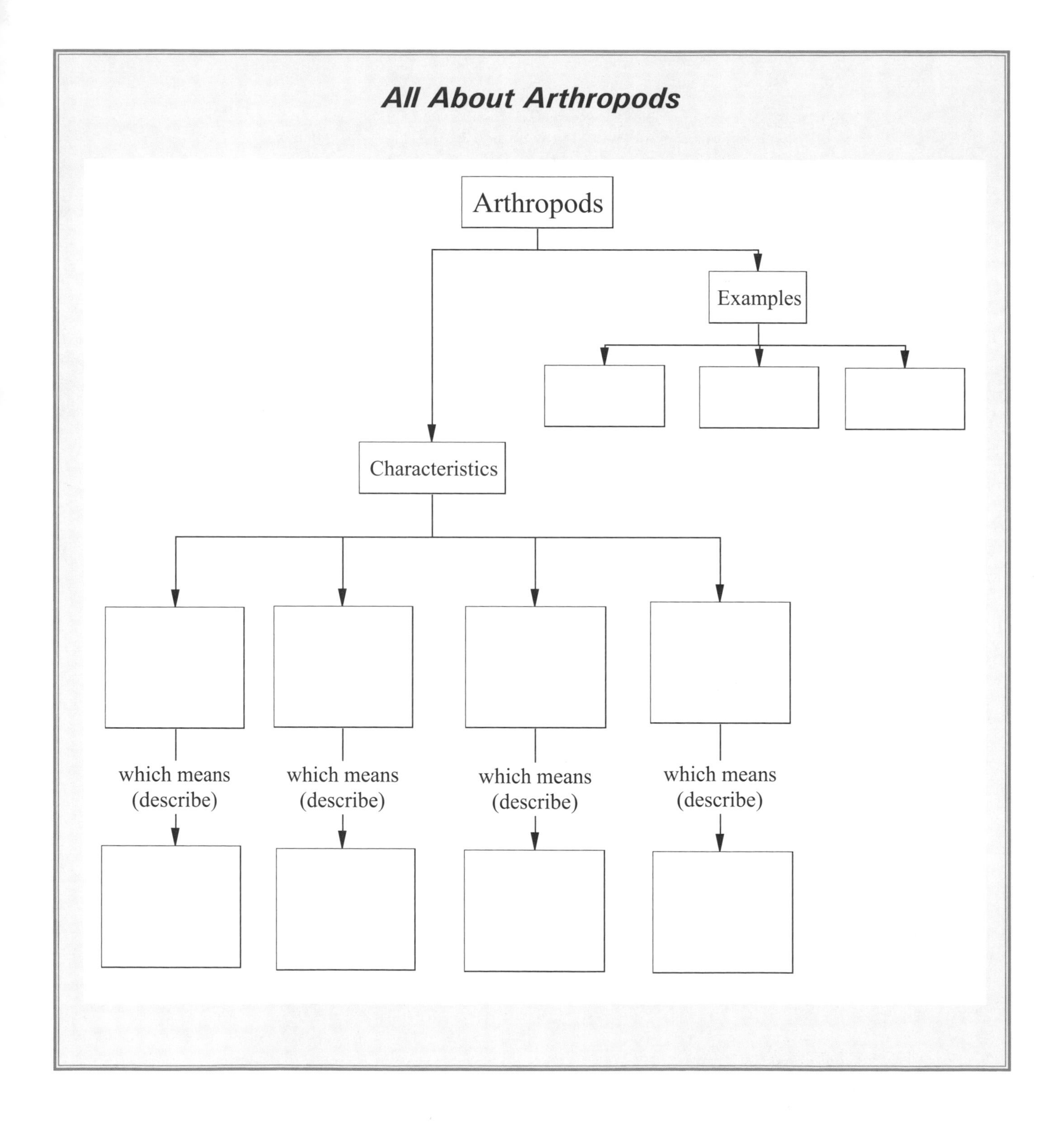
All About Arthropods
Arthropods
Examples
Characteristics
which means (describe)
which means (describe)
which means (describe)
which means (describe)

Chapter 8

The Case of the Tree Hit Man

Topics

- Plant structure and function (roots, stems, and leaves)
- Vascular tissue in plants (xylem and phloem)

NSES Content Standards (For Grades 5–8, Life Science)

- Living systems at all levels of organization demonstrate the complementary nature of structure and function. Important levels of organization for structure and function include cells, organs, tissues, organ systems, whole organisms, and ecosystems.
- Specialized cells perform specialized functions in multicellular organisms. Groups of specialized cells cooperate to form a tissue, such as a muscle. Different tissues are in turn grouped together to form larger functional units, called organs. Each type of cell, tissue, and organ has a distinct structure and set of functions that serve the organism as a whole. (NRC 1996, p. 156)

Reading Strategy

Previewing diagrams and illustrations

Background

The topic of plants does not excite many middle school students. Learning about the inner workings of xylem and phloem feels far removed from any practical applications in their lives. This chapter uses a (true life!) crime scenario to frame the study of water and sugar movement in plants.

First, students will do a few basic explorations to make sure that they have an underlying understanding of water and sugar movement in plants. Although it may seem like students should know these basic ideas, many students have never observed plants closely and are confused by these topics. However, once students observe water movement in a plant, they will be ready to learn about xylem and phloem.

Materials

- 4 small plants per class, such as geraniums, bean seedlings, pansies, or young radishes
- Dried potting soil (see Teaching Note on p. 73)
- Plastic wrap or plastic bags (or other materials requested by students)
- Watering can or spray bottle
- Celery stalks, preferably with leaves attached
- Food coloring
- Plastic cups (1 per group, plus 4 per class to use as pots for the mini-experiment)
- Knives to trim the celery (Note: Some schools do not allow students to use knives. If this is the case, the teacher should prepare the celery in advance.)
- Indirectly vented chemical splash goggles, aprons, vinyl gloves

SAFETY ALERT!

- The teacher will demonstrate safe procedures for using a knife.
- Point and push knife away from your body.
- Never use excess force when working with a knife.
- Wipe up any water on the floor when using a spray bottle, as it can be a slip or fall hazard.

Student Pages

- "The Case of the Tree Hit Man"
- Plant Police Academy
- Structure and Function in Plants

Exploration/Pre-Reading

In this exploration, students will begin with a mini-experiment to decide if roots or leaves absorb water for plants. Then they will observe the movement of water through a celery stem and compare how water and sugar move through a plant. These explorations require several days to complete, so you will need to plan ahead.

Begin by telling students that a terrible crime has occurred. Over the next few days, they will be participating in Plant Police Academy to learn more about it. As their first task, students need to design an experiment that will show which part of the plant absorbs water. Show students the four

Topic: What Are the Parts of a Plant?
Go to: *www.scilinks.org*
Code: LSB012

Topic: Plant Growth
Go to: *www.scilinks.org*
Code: LSB013

Two Possible Designs for the Mini-Experiment

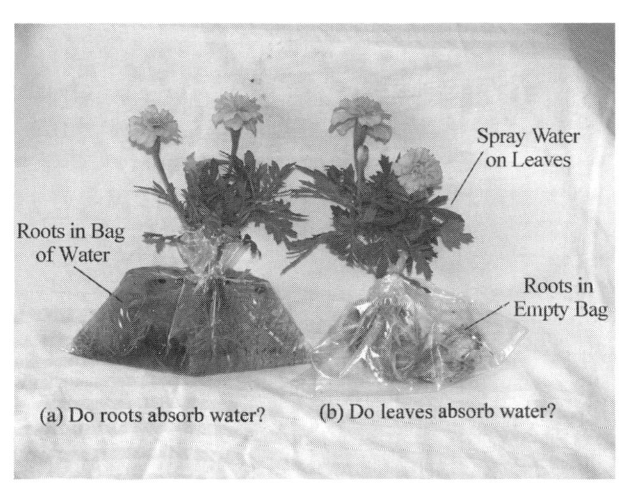

plants for their class, and ask the class to design an experiment that would let them show whether leaves or roots absorb the water. After a little brainstorming, most classes will come up with something similar to one of the two experimental designs shown in the figure above. (Note that each picture shows only one plant in each condition; your classes will have two plants in each group.)

Give students the handout Plant Police Academy, and have them draw the class's experimental design. Ask, "What happens to a plant when it runs out of water?" and make sure students understand that the plant will

TEACHING NOTE

If you purchase mature plants that have been well watered at the nursery, it can take up to three weeks for the plant with covered roots to wilt. You can shorten the time by using potting soil that has been thoroughly dried before planting. Place the soil on a cookie sheet and bake it at 250 degrees for two to four hours to remove the water. When you are ready to start the experiment, gently loosen the existing soil from the roots of your plants, rinse the remaining soil away with water, and then plant them in the dry soil. The plant with covered roots should then wilt in just two to four days.

wilt. They should then be able to fill in the prediction section for each experimental group.

The second activity, allowing celery to soak up colored water, may be familiar to students. This version, though, asks them to pay careful attention to where the color change takes place. The process requires at least one hour but works best when allowed to sit overnight.

Introduce the Reading. Tell students that they are now ready to read and investigate the crime. Give out the "The Case of the Tree Hit Man," and lead students to study the diagrams as described in the reading strategy section below.

Reading Strategy: Previewing Diagrams and Illustrations

Note that this strategy was first introduced in Chapter 6. If you have not used Chapter 6 with your class, tell students that in some books that they read, the pictures are extras. In science writing, however, the pictures and diagrams often carry a lot of important information. Looking at the pictures and making predictions about what they mean before reading can help make the text easier to understand.

If you have already introduced this strategy, remind students about the importance of pictures in science text. Then tell them that they are going to practice three questions that they can use to help them preview diagrams and illustrations.

Place students into their reading groups, and direct them to look at Figure 1 (p. 76). The Leaders should describe what they see in the diagram, without worrying about whether they know the correct terms. Then the Flag Flyers should predict what the diagram illustrates. Finally, the Interpreters should come up with at least one question about the diagram that might be answered in the text. Have one group share its responses with the class.

Continue to Figure 2 (p. 77), but this time the Interpreter should describe the diagram, the Leader should make the prediction, and the Flag Flyer should come up with a question. Continue this pattern until the students have discussed each diagram. Then have them proceed to reading the text as usual.

Journal Questions

When you looked at the diagrams before reading, your group discussed three questions:

- What do you see in the diagram?
- What might the diagram be illustrating?
- What question do you think the text will answer about the diagram?

Which of these three questions was most helpful for understanding the text? Why?

Application/Post-Reading

- Graphic Organizer: Structure and Function in Plants
- Pulling It Together in Writing: Give students the following prompt: *The town of Magnolia Springs has decided to prosecute the tree hit man. You are the police detective called in to explain the case to the jury. Explain how the hit man killed the tree, and include a diagram to help them understand.*
- Pulling-It-Together Focus Point: Phloem cells carry food and sugars through the plant. In a tree, the cells are located in a ring just beneath the bark. The tree hit man cut through the phloem cells and, therefore, starved the base of the tree.

References

Associated Press. *Birmingham News*. 1993. Giant oak from acorn of Columbus' day found mostly dead in Magnolia Springs. May 20.

Ball, E. *Birmingham News*. 1991. Magnolia Springs' mighty oak winning fight for life. March 31.

Mitchell, G. Associated Press. 2000. Visitors find inspiration: Remains of 500-year-old tree still park's centerpiece. November 19. *http://www2.ljworld.com/news/2000/nov/19/visitors_find_inspiration*

National Resource Council (NRC). 1996. *National science education standards*. Washington, DC: National Academies Press.

Save the Tree Committee. n.d. Inspiration Oak vital statistics (handout provided at Inspiration Oak Park).

Save the Tree Committee. n.d. Answers to frequently asked questions (handout provided at Inspiration Oak Park).

REMEMBER YOUR CODES

! This is important.

✓ I knew that.

X This is different from what I thought.

? I don't understand.

The Case of the Tree Hit Man

The assassin crept across the lawn in the dead of night. He worked quietly and efficiently. There were only a few hours until morning, and his work would need to be completed by the time the sun came up. He had been hired to kill one of Magnolia Springs' oldest and most respected residents: a 500-year-old oak tree.

The tree, named Inspiration Oak, had been growing since before the time of Christopher Columbus. It had survived the birth of the small town and the invention of the automobile. By 1990, it stood 65 feet tall, with a trunk that was almost 30 feet in diameter.

The owner of the land had decided to chop down the tree and build a gas station. The people of Magnolia Springs loved that tree, so county officials denied the permit for a gas station and collected money to buy the land instead.

The owner's plans were foiled, and all because of that tree. But what if something happened to the tree? The owner got on the phone and called a hit man.

It's not easy to kill a big tree in a hurry. Chopping it down would take time and be noisy. Neighbors would be sure to notice. Poison might work, but it would take massive amounts of poison and could take years to finish the job. But the tree hit man had a plan because he knew how trees were organized.

What the Hit Man Knew

Trees, like most plants, have three main parts: the roots, leaves, and stem. The roots hold the tree firmly in the ground so that even a strong wind will not knock it over. They grow deep into the soil and can stretch for hundreds of feet in all directions. As shown in Figure 1, the root network is as big as the trunk and the branches combined. Trees need this vast network of roots to absorb water and minerals for growth.

At the other end of a tree, you find the leaves. Leaves make food for the plant in a process called *photosynthesis*. The leaves get carbon dioxide from the air and water from the roots. Then they use energy from sunlight to convert the water and carbon dioxide into sugars.

The stem, or tree trunk, holds the leaves high in the air so they can get enough sunlight. The trunk also connects the roots and leaves. Inside the trunk, tubes carry water and minerals from the roots up to the

Figure 1. Tree Roots

Roots make up about half of the total size of a tree.

Figure 2. One Way That Vascular Bundles Can Be Arranged in a Stem

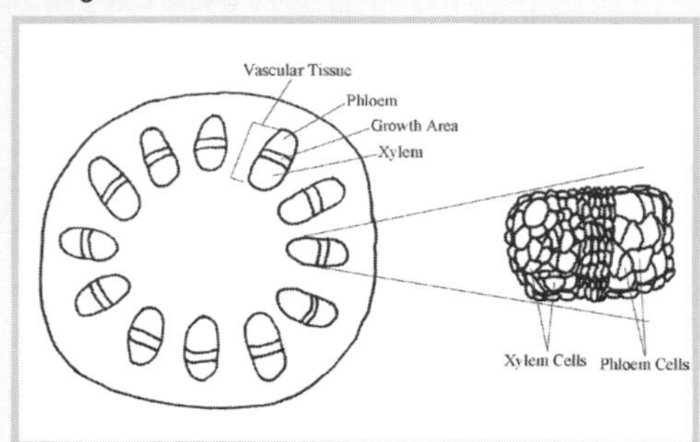

leaves. Another set of tubes carries sugars from the leaves to the roots.

These tubes are made of specialized plant cells that connect end to end. The cells that carry water up are called *xylem*. The cells that carry sugar down are called *phloem*. Together, the xylem and phloem are called the *vascular tissue*. Vascular tissue carries water and sugars throughout the roots, leaves, and stems.

In most plants, xylem and phloem are found in bundles throughout the stem. But in trees, the xylem grows in the center of the trunk and all of the phloem lie just below the bark on the outside.

Quick and Dirty Business

The hit man worked quickly to chisel into the tree. He made a perfect ring, six inches wide and six inches deep. With each cut, he removed the bark and the phloem layer. Thick, sticky liquid oozed from the wounds. Soon the tree had no way to get sugar from its leaves to its roots. It was only a matter of time until the root cells died of starvation.

In the morning, the townspeople were horrified. They brought in tree experts from around the country, who built an elaborate intensive care unit to try to reconnect the phloem tubes. But it was too late. The tree could not be saved.

No charges were ever brought in the case of the tree hit man, but the town of Magnolia Springs came together in its effort to save the tree. They went through with their plans to buy the land and build a park. It is a beautiful, quiet picnic spot that is marked with the giant stump of Inspiration Oak.

Figure 3. Xylem and Phloem in a Tree Trunk

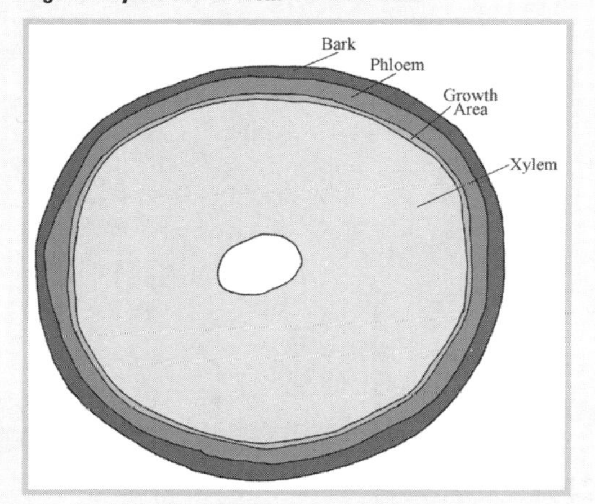

THE BIG QUESTION

How do water and food (sugars) move through a plant?

Plant Police Academy

A terrible crime has occurred. In a few days, you will learn more about it. But first, you need to come to plant police academy to find out more about plants.

Part 1: Which part of a plant absorbs water: roots or leaves? (mini-experiment)

Draw a diagram of your two experimental groups.

Prediction
If roots absorb water, this plant will (circle one): wilt not wilt.

If leaves absorb water, this plant will (circle one): wilt not wilt.

Results: Did the plant wilt? ________

Prediction
If roots absorb water, this plant will (circle one): wilt not wilt.

If leaves absorb water, this plant will (circle one): wilt not wilt.

Results: Did the plant wilt? ________

Claims and Evidence: Which part of the plant absorbs water? What evidence do you have from your experiment?

Part 2: Water Movement in the Stem of a Plant

Fill a cup halfway with water and add 8 to 10 drops of food coloring. Get a stalk of celery from your teacher. Cut about 2 cm from the bottom and top of the celery stalk to get rid of any dried parts. Then place the stalk in the colored water, leaf side up. Let it sit in the water for at least an hour.

a. What happened to the celery?

b. Based on these results, does water move up or down the stem of the celery plant?

c. Chop the celery stalk in half and look at the inside with a hand lens.

   Draw what you see here, and label the areas where you can see food coloring.

d. Did the water move everywhere in the celery, or just in certain places?

Part 3: How Water and Sugars Move in a Plant

Sugar is made in the leaves of a plant and then travels to all the cells of the plant. In the diagram to the right, the arrow shows the direction that sugar moves.

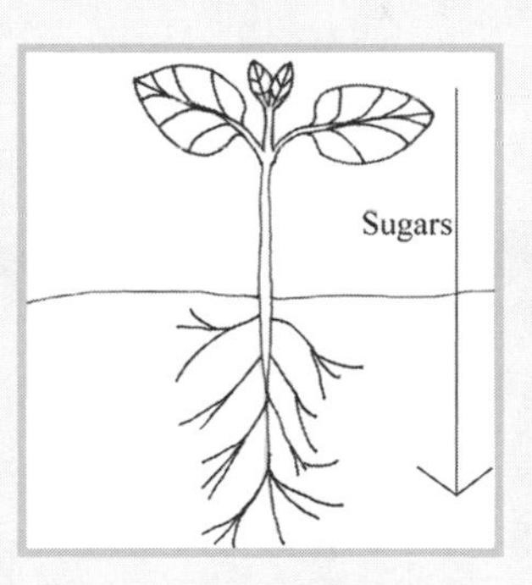

Draw a second arrow to show the direction that water moves in the plant.

Think about the direction of your arrows. Can food and water move in the same tubes inside the plant stem? Why or why not?

Congratulations! You have completed the Plant Police Academy. Now it is time to find out more about the Case of the Tree Hit Man.

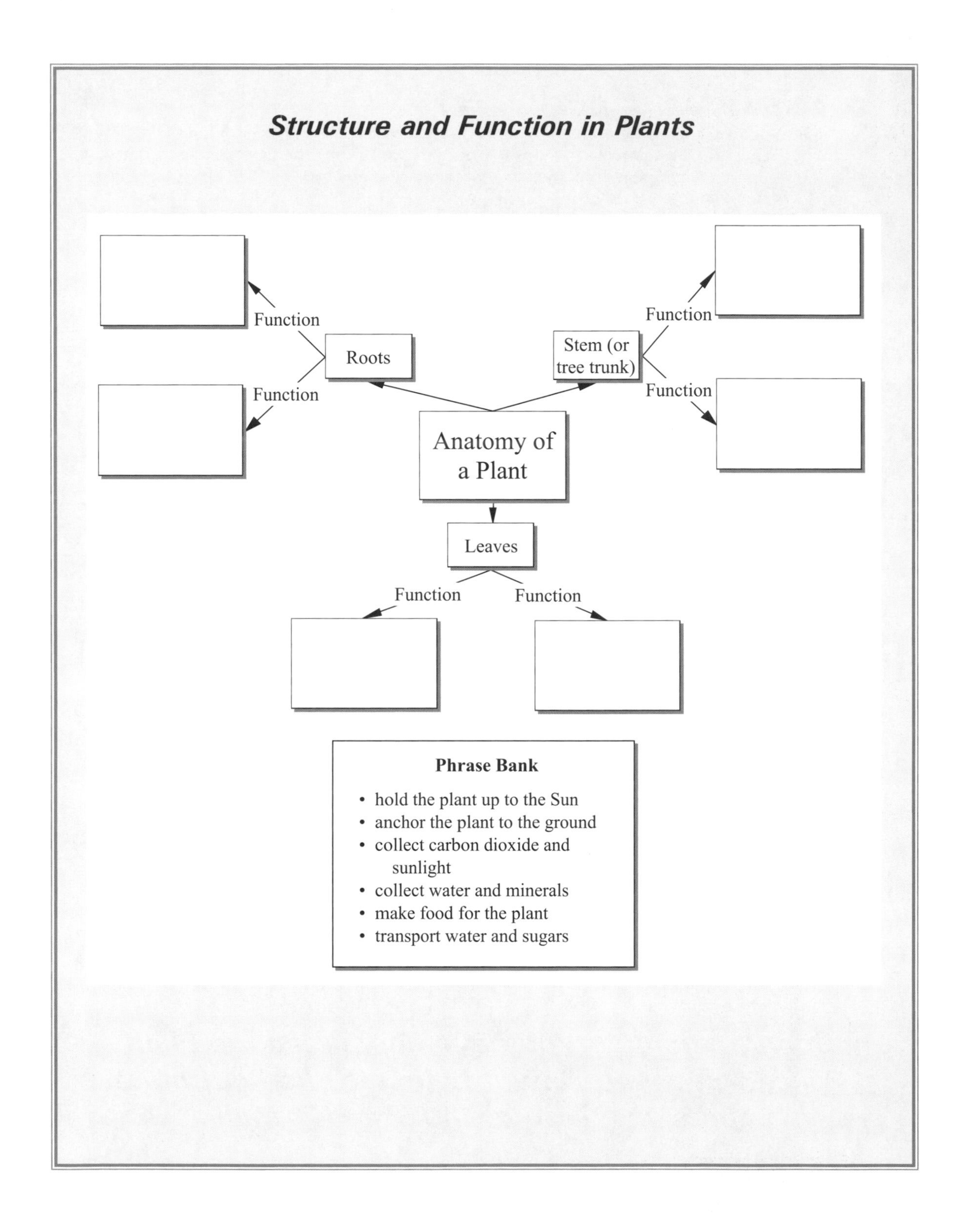
Structure and Function in Plants
Function
Roots
Function
Function
Stem (or tree trunk)
Function
Anatomy of a Plant
Leaves
Function
Function
Phrase Bank
• hold the plant up to the Sun
• anchor the plant to the ground
• collect carbon dioxide and sunlight
• collect water and minerals
• make food for the plant
• transport water and sugars

Chapter 9

A Gene for Drunkenness?

Topics

- Genetics
- Gene and environment interactions
- Human disease

NSES Content Standards (For Grades 5–8, Life Science)

- Every organism requires a set of instructions for specifying its traits. Heredity is the passage of these instructions from one generation to another.
- Hereditary information is contained in genes, located in the chromosomes of each cell. Each gene carries a single unit of information. An inherited trait of an individual can be determined by one or by many genes, and a single gene can influence more than one trait. A human cell contains many thousands of different genes.
- The characteristics of an organism can be described in terms of a combination of traits. Some traits are inherited and others result from interactions with the environment.
- Disease is a breakdown in structures or functions of an organism. Some diseases are the result of intrinsic failures of the system. Others are the result of damage by infection by other organisms. (NRC 1996, p. 157)

Reading Strategy

Chunking

Background

The genetics of some human traits, such as tongue rolling or sickle cell anemia, are relatively straightforward. A person carries a gene that codes for a protein and creates a phenotype. However, over the past few decades, there has been an explosion of research into more complicated scenarios in which both genetics and environmental causes play a part in determining a trait. Alcoholism is a well-researched example of this complexity. It is also a timely topic for adolescents, as many alcoholics begin drinking during their middle school years.

For years, counselors have known that certain environmental factors, such as abuse or a family atmosphere that encouraged drinking, could put people at risk for alcoholism. In the past few decades, scientists have been uncovering genes that also make people prone to alcoholism. The interactions between the environmental and genetic influences on this behavior make it difficult to say exactly what "causes" alcoholism.

Examining the relationship between genes and alcoholism can help students make wise choices about drinking, and it also can help them gain a wider perspective on genetics—a perspective that can help them understand a host of problems, from heart disease to obesity to depression.

As you teach this topic, you may want to watch for two misconceptions. First, there is a misconception that *environmental* factors refer just to conditions in a person's environment, such as they might study in an ecology unit. In this case, environmental factors refer to anything that is not genetic. These factors may or may not be under a person's control.

Second, there is also a misconception that if a person carries a gene that is linked to a specific trait, they are destined to develop that phenotype. Although that may be true for some conditions, such as Huntington's disease, it is not true for many others. Students need to know that in these cases, genes influence rather than determine a person's behavior.

SCILINKS
THE WORLD'S A CLICK AWAY

Topic: Genes and Traits
Go to: *www.scilinks.org*
Code: LSB014

Topic: Alcoholism
Go to: *www.scilinks.org*
Code: LSB015

Topic: Genetic Diseases, Screening, Counseling
Go to: *www.scilinks.org*
Code: LSB016

Topic: Huntington Disease
Go to: *www.scilinks.org*
Code: LSB017

Materials (all can be re-used across classes)

- 8 game pieces for each group of 2 or 3 students (Any place holder, such as squares of construction paper, would work.)
- 8 chips for each group of 2 or 3 students (For visual effect, these should have some thickness, such as pennies or poker chips.)
- 1 die for each group of 2 or 3 students (Packages of dice are available inexpensively at party supply stores.)
- 1 How Do the Chips Stack Up? Game Board for each student group (To make the board, copy the two pages and then tape them together.)

Student Pages

- "A Gene for Drunkenness?"
- How Do the Chips Stack Up? Game Board
- How Do the Chips Stack Up? Questions to Consider
- Causes of Alcoholism in Teens and Young Adults

Exploration/Pre-Reading

In this exploration, students will roll a die to create a patient profile and look at the effects of environmental and genetic factors on that patient's risk of alcoholism. Then they will decide what advice to give the patient.

Begin by asking students to recall a disease that you have studied that is caused by genetics (such as sickle cell anemia, Huntington's disease, or hemophilia). Explain that these types of health problems are unusual; most traits are caused by a combination of genetics and other factors.

Ask students to consider addiction to alcohol (alcoholism). Ask what kinds of factors might lead a person to become an alcoholic. Most students will list environmental factors, such as drinking too much, depression, or having something really bad happen. Explain that these are considered *environmental* factors. You will need to take a moment to explain that when the word *environment* is used in genetics, it means more than just the grass and trees and other things we think of in an ecosystem. In genetics, *environment* refers to anything other than genes.

Tell students that they will be doctors for the day. A patient is coming to talk about his or her risk of alcoholism. First, students will need to find out what environmental and genetic risks their patient faces.

Give each group of two or three students a "How do the Chips Stack Up?" game board, eight game pieces, eight chips, and one die. Have students place a game piece on each starting square. They should follow the directions at the top of game board to roll the die to determine the patient's risk for each factor, then place a chip on the appropriate stack. After rolling for all eight factors, the students can look at the stacks for a visual representation of the patient's risk.

Note that this simulation is a simplification of the relationship between risk factors and alcoholism. However, having students tally risk factors, rather than trying to understand complicated mathematical models, allows them to explore the general concept of how a variety of factors can affect the expression of genes.

Hand out How Do the Chips Stack Up? Questions to Consider, give students time to answer the questions, and then have a short class discussion. Allow students to share their observations, particularly about risk factors that may have surprised them. Then follow up with Question 5, about what a person can do to avoid becoming an alcoholic. Make sure students observe that even someone at high risk can avoid alcoholism if they never begin drinking. This is an important point to make with your class. Students often have a hard time understanding the difference between risk and fate. You may have several children of alcoholics in your classes. It is valuable for these students to know that they are at increased risk, but they also need the reassurance that they have control over their futures.

As a teacher, be aware that this may be a sensitive topic for some of your students, especially if they come from families with a history of alcoholism. A lot of research indicates that these children are likely to repeat the family pattern if they don't have help. Although understanding risk factors can be important, it is generally insufficient. If a student comes to you and mentions that he or she is living with an alcoholic, try to arrange for the student to speak to the school counselor or another professional.

Introduce the Reading. Tell students that the text they are about to read will tell them more about the genetic and environmental causes of alcoholism.

Reading Strategy: Chunking

To introduce this strategy, place the following sentence on the board:

> *Counselors and others who work with people who have alcohol addictions have known for a long time that people who have certain types of experiences often become alcoholics.*

Point out that this sentence, like many sentences in science writing, has a lot of ideas crammed together. It would be difficult to try to understand all of the ideas at one time, but if students break the sentence into chunks, they can think about each piece individually.

Add slashes (/) to the sentence on the board so that it reads like this:

> *Counselors and / others who work with people who have alcohol addictions / have known for a long time / that people who have certain types of experiences / often become alcoholics.*

Talk them through the sentence, one section at a time. Start with the word *counselors*. Ask, "Why would an alcoholic go to a counselor?" (to get help to stop drinking; to talk about his or her problems) Then look at the next phrase: *others who work with people who have alcohol addictions*. Ask, "Who do you think the *others* mentioned in this sentence are?" (doctors, researchers, ministers, etc.)

The next phrase indicates that all of these people (counselors, doctors, etc.) have known something. Ask students what these people have known. Finally, point out that the phrase *certain types of experiences* is vague. Have students predict what the text might tell them in the next few sentences (what those experiences are).

Explain that chunking a sentence is like eating a pie. People cannot put the whole pie in their mouth at one time; everyone eats it bite by bite. Like eating, some people will take bigger bites than others. That means that some people will need to break a sentence into more chunks than others, and that is okay. For this article, students can separate the chunks using slashes, like you did on the board. When they are reading something they aren't allowed to write on, they can chunk it in their head or cover up the parts of the sentence they aren't thinking about.

Journal Question

Chunking is especially useful when you are reading long sentences full of new information. Think of a topic you know a lot about. Imagine that you are writing for someone who knows very little about that topic. Write a sentence that gives a lot of information on your topic. Use slashes to mark how your reader might chunk that sentence.

Application/Post-Reading

- Graphic Organizer: Causes of Alcoholism in Teens and Young Adults
- Pulling It Together in Writing: Alcoholism can serve as a model for thinking about a number of different diseases. Give students the following writing prompt to see how well your students can extend the ideas that they have learned in this chapter:

Heart attacks are also caused by a combination of environment and genes. On the environment side, risk factors include

- *lack of exercise*
- *eating foods that are high in fat and cholesterol*
- *smoking*

On the genetics side, there are genes that make it more likely that a person will have a heart attack. Your friend says that because his dad and grandpa both had heart attacks, he knows he will have one, too. What would you say to him?

- Pulling-It-Together Focus Point: Genetics can increase the friend's risk of a heart attack, but they do not ensure that one will occur. The friend can focus on environmental factors that are within his control (such as exercise and eating habits) to reduce his risk.

References

Black, C. 2001. *It will never happen to me: Growing up with addiction as youngsters, adolescents, and adults.* 2nd ed. Center City, MN: Hazelden.

Dick, D., and A. Agrawal. 2008. The genetics of alcohol and other drug dependence. *Alcohol Research and Health* 31 (2): 111–118.

Ducci, F., and D. Goldman. 2008. Genetic approaches to addiction: Genes and alcohol. *Addiction* 103 (9): 1414–1428.

National Heart, Lung, and Blood Institute. 2008. Who is at risk for a heart attack? *www.nhlbi.nih.gov/health/dci/Diseases/HeartAttack/heartattack_risk.html*

National Institute of Alcohol Abuse and Alcoholism. 1997. Youth drinking: Risk factors and consequences. *Alcohol Alert* 37 (July). *http://pubs.niaaa.nih.gov/publications/aa37.htm*

National Resource Council (NRC). 1996. *National science education standards.* Washington, DC: National Academies Press.

REMEMBER YOUR CODES

! This is important.

✓ I knew that.

X This is different from what I thought.

? I don't understand.

A Gene for Drunkenness?

Michelle had her first beer when she was 12. She grabbed a couple from the fridge at home and invited her best friend to drink them with her. By the time she was in high school, she was drinking every day. On the weekends, she would go to parties and drink until she passed out. One morning, she couldn't even remember whose party she'd gone to the night before or how she had gotten home. She suddenly realized that she was addicted to alcohol. How is it that some people safely drink a little alcohol now and then, but others, like Michelle, begin drinking and find themselves unable to stop?

The answer lies in a combination of a person's environment, genes, and choices. Some health problems, such as sickle cell anemia, are caused entirely by genes. Other health problems are caused by environmental factors, such as a broken leg after being hit by a car. But many health problems come from the environment and genetics working together. Alcoholism is one of these diseases. Scientists have identified several factors that put a person at risk of becoming an alcoholic.

Environmental Factors

Counselors and others who work with people who have alcohol addictions have known for a long time that people who have certain types of experiences often become alcoholics. Children who grow up with a parent who is an alcoholic are more likely to become alcoholics. Even if that parent isn't biologically related, such as a stepparent or an adopted parent, the child is still at risk of becoming an alcoholic. Childhood trauma, such as child abuse or neglect, can also lead to alcoholism.

____

Friends affect a person's chances of becoming an alcoholic as well. Someone who hangs out with heavy drinkers is likely to drink with them. Friends may also encourage a young person to start drinking before it is legal. Kids who start drinking when they are younger than 15 are four times more likely to become alcoholics than people who don't start drinking until they are at least 21.

Genetic Factors

Genetics also have a role in determining if a person is at risk for alcoholism. A lot of different genes are involved, and scientists are still figuring out how all of those genes work. Many of the genes being studied are important in the processes described below.

The Flushing Reaction

Some people feel very sick when they drink. Shortly after a person starts drinking, his or her face turns red, or flushes. The person gets a headache, his or her heart starts beating quickly, and the person feels like throwing up. This group of symptoms is called the *flushing reaction*. People who have the flushing reaction do not enjoy drinking, and they are unlikely to become alcoholics. On the flip side, those who can "hold their liquor"—who can drink a lot before they start to feel bad—are much more likely to become alcoholics. Genes control whether a person experiences the flushing reaction when they drink.

Brain Reception

When a person drinks, much of the alcohol ends up in the brain, where it interacts with brain cells. Because of genetics, some people's brain cells have a more powerful reaction to alcohol than others. This strong reaction makes them feel relaxed when they drink. It also means that they become addicted more easily and have intense withdrawal symptoms.

Response to Trauma

Many people who were abused as children struggle with anger, depression, and anxiety. They have a greater risk of becoming alcoholics. But some people who were abused seem to be able to move on more easily and are not as likely to become alcoholics. Scientists have found that many of these people share a set of genes.

Scientists are not sure exactly how these genes work, but they may have something to do with how the brain stores memories. They may also affect the chemicals that the body releases in response to stress. These genes interact directly with environmental influences. They make it less likely that a person who had severe trauma as a child will become an alcoholic.

Choices

By studying patterns in families, scientists believe that about 50% of a person's risk of becoming an alcoholic is genetic. The rest is environmental. But, of course, all of these factors describe *risk*. They don't say for sure what will happen. Even a person with no obvious risk factors can become an alcoholic. Likewise, a person with many risk factors can avoid becoming an alcoholic. When people know the risks, they can make careful decisions about whether and when to drink.

THE BIG QUESTION

The title of this article asks, "A Gene for Drunkenness?" Is there a gene for drunkenness? How would you describe what causes alcoholism?

How Do the Chips Stack Up? Game Board

Roll the die and move one of your game pieces the number of spaces shown on the die to find out what risk factors affect your patient. Follow the directions in that square to add a chip to the high-risk or low-risk stack. After you have rolled for all eight risk factors, compare the two stacks to see how the chips stack up.

1. The patient's parents ...	Offered their children alcohol *Add 1 to high risk.*	Warned about the dangers of alcohol *Add 1 to low risk.*	Offered their children alcohol *Add 1 to high risk.*	Warned about the dangers of alcohol *Add 1 to low risk.*	Offered their children alcohol *Add 1 to high risk.*	Warned about the dangers of alcohol *Add 1 to low risk.*
2. Did the patient have an alcoholic parent?	Yes *Add 1 to high risk.*	No *Add 1 to low risk.*	Yes *Add 1 to high risk.*	No *Add 1 to low risk.*	Yes *Add 1 to high risk.*	No *Add 1 to low risk.*
3. The patient's friends ...	Drink heavily *Add 1 to high risk.*	Do not drink *Add 1 to low risk.*	Drink heavily *Add 1 to high risk.*	Do not drink *Add 1 to low risk.*	Drink heavily *Add 1 to high risk.*	Do not drink *Add 1 to low risk.*
4. When did the patient have his or her first drink?	Younger than age 15 *Add 2 to high risk.*	Between 15 and 20 *Add 1 to high risk.*	Age 21 or older *Add 1 to low risk.*	Younger than age 15 *Add 2 to high risk.*	Between 15 and 20 *Add 1 to high risk.*	Age 21 or older *Add 1 to low risk.*
5. Was the patient abused as a child?	Yes *Add 1 to high risk.*	No *Add 1 to low risk.*	Yes *Add 1 to high risk.*	No *Add 1 to low risk.*	Yes *Add 1 to high risk.*	No *Add 1 to low risk.*

6. Gene Set 1—Flushing Reaction	Can drink a lot of alcohol before feeling sick *Add 1 to high risk.*	Feels sick after several drinks *Don't add any chips.*	Feels sick when drinking even a little alcohol *Add 1 to low risk.*	Can drink a lot of alcohol before feeling sick *Add 1 to high risk.*	Feels sick after several drinks *Don't add any chips.*	Feels sick when drinking even a little alcohol *Add 1 to low risk.*
7. Gene Set 2—Relaxation Response	Alcohol causes strong feelings of relaxation *Add 1 to high risk.*	Alcohol causes normal feelings of relaxation *Don't add any chips.*	Alcohol causes strong feelings of relaxation *Add 1 to high risk.*	Alcohol causes normal feelings of relaxation *Don't add any chips.*	Alcohol causes strong feelings of relaxation *Add 1 to high risk.*	Alcohol causes normal feelings of relaxation *Don't add any chips.*
8. Gene Set 3—Response to Stress	Patient has genes that cause the body to have negative reactions to stress *Add 1 to high risk ONLY IF the patient had yes for #5.*	Patient has genes that help the body overcome stressful situations *Add 1 to low risk ONLY IF the patient had yes for #5.*	Patient has genes that cause the body to have negative reactions to stress *Add 1 to high risk ONLY IF the patient had yes for #5.*	Patient has genes that help the body overcome stressful situations *Add 1 to low risk ONLY IF the patient had yes for #5.*	Patient has genes that cause the body to have negative reactions to stress *Add 1 to high risk ONLY IF the patient had yes for #5.*	Patient has genes that help the body overcome stressful situations *Add 1 to low risk ONLY IF the patient had no for #5.*

High Risk

Low Risk

How Do the Chips Stack Up? Questions to Consider

1. Which stack was taller for your patient, the high risk or the low risk?

2. Are there ways your patient could decrease his or her risk? How?

3. What advice would you give your patient about drinking?

4. Did any of the risk factors surprise you? Why or why not?

5. Risk only tells part of the story. What can your patient do to be sure that he or she does not become an alcoholic, regardless of his or her risk?

Causes of Alcoholism in Teens and Young Adults

Risk Factor	Genes or Environment?	Description (Tell more about the risk factor.)	Increases or Decreases Risk?
Alcoholic parent			
Trauma			
Friends			
Age of first drink			
Flushing reaction			
Brain reception			
Response to trauma			

Chapter 10

Oh! I Gotta Pee!

Topics

- Urinary system
- Homeostasis

NSES Content Standards (For Grades 5–8, Life Science)

- The human organism has systems for digestion, respiration, reproduction, circulation, excretion, movement, control, and coordination, and for protection from disease. These systems interact with one another.
- All organisms must be able to obtain and use resources, grow, reproduce, and maintain stable internal conditions while living in a constantly changing external environment.
- Regulation of an organism's internal environment involves sensing the internal environment and changing physiological activities to keep conditions within the range required to survive. (NRC 1996, pp. 156–157)

Reading Strategy

Pause, retell, and compare

Background

SCILINKS
THE WORLD'S A CLICK AWAY
Topic: Urinary System
Go to: *www.scilinks.org*
Code: LSB018

If you ask a classroom of middle-school students to trace the path of a drink from their mouth to the toilet, you can uncover many misconceptions about the urinary system. Many students will call upon a hypothetical tube running from the stomach or intestines to the bladder. This misconception makes it hard for them to understand the critical role of the kidneys. Therefore, this chapter starts out by having students identify their current ideas about urination and leads them to correct any misconceptions.

Materials

SAFETY ALERT!
- Use appropriate ventilation to eliminate preservative fumes.
- No food or drink is allowed during work with dissected mammals.
- Wash hands with soap and water after working with dissected mammal.

- Variety of diagrams of the human abdominal cavity (from textbooks, internet sources, posters, college dissection guides, library books, etc.)
- 1 or more 3-D models of the human body (optional)
- Predissected pig, rat, or other mammal (optional)
- Indirectly vented chemical splash goggles, aprons, gloves (vinyl) (for dissected mammals)

Student Pages

- "Oh! I Gotta Pee!"
- What to Do in Your Reading Groups: Practicing Pause, Retell, and Compare
- The Liquid Balancing Act

Exploration/Pre-Reading

In this exploration, students will look at diagrams, models, and possibly animal dissections to trace the path of urine through the body. Ideally, this exploration will be completed during (or after) dissecting a mammal, such as a rat or fetal pig. However, it can still be done effectively using only diagrams and models. Before class, gather a variety of diagrams of the human internal organs. Make sure you have at least one diagram for each group of students. In addition, if your school has one or more models of the human body, make those available to your students. Even if your class does not actually dissect an animal, if you have access to a predissected mammal so that students could consult the organs of the abdominal cavity, this would be helpful.

Place your students in groups of two or three (in their reading groups, if you are using those), and present this scenario and question:

> *You just drank a large glass of water. Later on, you have to pee. Where does that liquid go from the time it enters your body until it reaches the toilet?*

Tell students they have 15 minutes to try to solve the problem. They can answer in a list or a diagram, and they may consult any of the materials you have provided. To get students to engage the visual information as much as possible, you may need to make a rule that they cannot look up the answer in their books (although books rarely address this question in a straightforward manner).

Your interaction with the groups is critical during this process. Most students, if they have studied the digestive system, can get the liquid as far as the stomach or intestines. Some groups will become stuck at this point. You might suggest to these groups that they start thinking backward from where the liquid exits. Others will suggest a hypothetical tube by which the water could move to the bladder. Ask these students to find evidence of such a tube in the diagrams, models, or dissected animal. As students zero in on the kidneys as a key part of the system, ask them to look for what "tubes" might carry liquid to the kidneys. They will see that only blood vessels connect to the kidneys. Groups do not have to arrive at a correct answer, as long as they have made two important discoveries: recognizing that there is no direct connection from the digestive system to the bladder, and seeing that only blood vessels connect to the kidneys. Indeed, students will be more engaged in the article if you refuse to tell them whether or not their answer was correct until after they read it.

Introduce the Reading. Tell students that the text they will read will clarify the problem they have been working on and allow them to check their answers.

Reading Strategy: Pause, Retell, and Compare

Write the following paragraphs from the reading on the board and have a student read them aloud:

After you ate lunch, your food and drink started the long journey through your digestive system to your intestines. Blood vessels surrounding your intestines absorb the water and nutrients into your bloodstream and carry them to all of the cells of your body.

As your cells use the nutrients in chemical reactions, they produce leftover material. Some of these leftovers are poisons, or toxins. The toxins and other waste move into the bloodstream and are carried to the kidneys.

Say "There was a lot of information in those two paragraphs. Sometimes I have a hard time remembering everything when I read a lot of new information at one time. One way to remember more is to pause and try to tell yourself what you just read."

Then model for the students how to use pause, retell, and compare. Tell them some of the things you remember from the text, but do not include all of the information. For example, you might say, "I remember that when I eat food, it goes to my intestines and to my body, and then there are some poisons." Then reread the passage aloud and ask the class what else you should remember. Tell students that by pausing, telling yourself the information, and comparing this information with what you just read, you will remember a lot more when you are done reading.

If students read individually, instruct them to pause after each section of the text to jot down what they remember, then look back in the text to add any important information to their list. If students read in groups, give them "What to Do in Your Reading Groups: Practicing Pause, Retell, and Compare," and go over the procedure. You may wish to have groups continue to use this modified procedure for several of the lessons that follow in this book, even while introducing other strategies.

Journal Questions

Some people think that good readers don't have to use a strategy such as pause, retell, and compare. In fact, most people who remember what they read use a strategy like this. Did you find it easier to remember what you read after trying this strategy? Why or why not?

Application/Post-Reading

- Graphic Organizer: The Liquid Balancing Act
- Pulling It Together in Writing: Kidneys are some of the most commonly transplanted organs, and the need for more donors is great. This need provides a chance for a real-world application of your students' new knowledge. Have students write a letter to their school or city newspaper describing the importance of kidney donation. Send the best letters to the newspaper. Alternatively, have students create "advertisements" for becoming an organ donor that can be posted around the school. Provide students with the following prompt:

Many health problems can damage the kidneys, including diabetes, high blood pressure, certain infections, and drug overdoses. Kidney failure, in which the kidneys stop working, can be deadly. Many people who experience kidney failure need a kidney transplant to survive. At the end of 2007, nearly 80,000 people in the United States were waiting for kidneys to become available for transplant. Write a letter to the newspaper to educate others about the need for people to donate their kidneys. Be sure to explain what the kidneys do in the body and why they are so important for survival.

- Pulling-It-Together Focus Point: Kidneys are important to survival because they clean toxins from the blood by removing them from the body through urine; they also control the amount of water in the body.

Topic: Organ Transplants
Go to: *www.scilinks.org*
Code: LSB019

Topic: Organs of Excretion
Go to: *www.scilinks.org*
Code: LSB020

References

National Kidney Foundation. 2009. Kidney disease: Causes. *www.kidney.org/atozItem.cfm?id=83*

National Resource Council (NRC). 1996. *National science education standards*. Washington, DC: National Academies Press.

Scientific Registry of Transplant Recipients. n.d. Waitlist and transplant activity for liver, kidney, heart, and lung, 1998–2007. *www.ustransplant.org/csr/current/fastfacts.aspx*

REMEMBER YOUR CODES

! This is important.
✓ I knew that.
X This is different from what I thought.
? I don't understand.

Oh! I Gotta Pee!

You guzzle a 20-ounce soda at lunch, and soon you have to pee. You wave your hand wildly in class. "I gotta go to the bathroom right away!" you tell your teacher.

Why is your body in such a hurry to get rid of the extra liquid?

Body in Balance

Your body is constantly working to keep everything in balance. For example, when you get too hot, you sweat to cool off. When you get too cold, you shiver to warm up. This process of keeping internal conditions stable is called *homeostasis*.

Your body works to maintain a balance of liquids, too. When you take a drink, you increase the amount of liquid in your body. To maintain homeostasis, you need to get rid of some extra liquid. You lose a little of it through sweat, some of it evaporates through your mouth when you exhale, and some goes right on out with your poop. But most of the liquid leaves your body in urine, or pee. When you pee, your urinary system puts this extra liquid to work carrying waste products out of your body.

Source: Sam Korn. *http://commons.wikimedia.org/wiki/File:Unbalanced_scales.png*

Making Waste

After you ate lunch, your food and drink started the long journey through your digestive system to your intestines. Blood vessels surrounding your intestines absorb the water and nutrients into your bloodstream and carry them to all of the cells of your body.

As your cells use the nutrients in chemical reactions, they produce leftover materials. Some of these leftovers are poisons, or toxins. The toxins and other waste move into the bloodstream and are carried to the kidneys.

Kidney Power

Your kidneys are like a wastewater treatment plant for your body. They filter and clean your blood. You have one kidney on

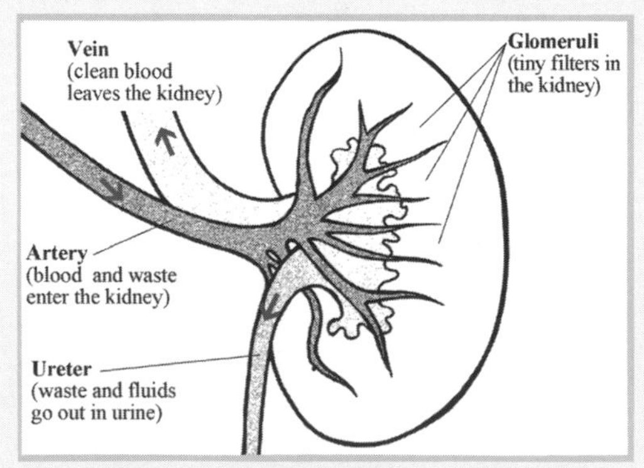

Source: National Institute of Diabetes and Digestive and Kidney Diseases, NIH. *www.catalog.niddk.nih.gov/ImageLibrary/detail.cfm?id=33*

each side of your body, located just below your ribcage against your back.

Blood comes to your kidneys through blood vessels called *renal arteries*. (Anything called *renal* has something to do with the kidneys.) Each kidney is made of about one million tiny filters. These filters, called *nephrons*, can only be seen through a microscope. Nephrons remove the excess water, toxins, and other waste products. This wastewater drains into a tube called the *ureter*. The clean blood returns to your body through the renal vein.

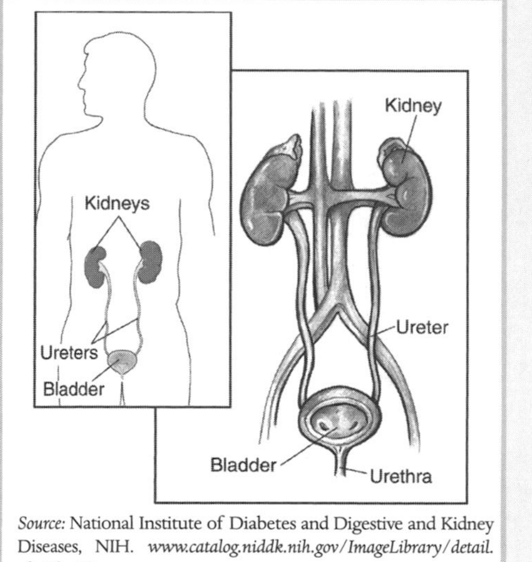

Source: National Institute of Diabetes and Digestive and Kidney Diseases, NIH. *www.catalog.niddk.nih.gov/ImageLibrary/detail.cfm?id=651*

Moving on Out

Urine is constantly dripping down the ureters into your bladder. The bladder is the body's storage tank for urine. When it collects a little less than a cup of urine, you feel the first urge to pee. If you don't go, it can stretch to hold almost four cups, but it is unhealthy to hold it that long.

When your teacher finally lets you go the bathroom, you relax a muscle that allows the urine to leave your bladder. The pee travels down another tube, called the *urethra*, to the outside of your body. Relief!

Color Codes

You have probably noticed that urine can be different colors. The color comes from the waste products that are dissolved in the water. These waste products include urea, which gives us the word *urine*; salts; and ammonia, which contributes to the smell. Another waste product is urochrome, which comes from the breakdown of old blood cells. Urochrome gives urine its yellow color.

If you drink a lot of liquids, your urine will be light because it will be mostly water. If you do not get enough to drink, your urine will be dark because your kidneys will get rid of only enough water to wash out the waste products. If your urine is dark, you may also feel thirsty. This is your body's signal that you need to drink more to maintain homeostasis.

The Urinary System

The urinary system—made of your kidneys, ureters, bladder, and urethra—plays a critical role in your survival. Needing to pee in the middle of class may be inconvenient, but urinating maintains the balance of liquids in your body and keeps your blood free of toxins.

THE BIG QUESTION

Look at your group's list or diagram of how water moves through your body from your mouth to the toilet. Does any of it need to be changed? If so, revise the list or diagram to show the correct pathway.

What to Do in Your Reading Groups: Practicing Pause, Retell, and Compare

1. The *interpreter* reads the Big Question aloud. Remember, this is what you are trying to learn!

2. Everyone reads the first section quietly and marks ✓, ?, !, and x while reading.

3. The *leader* tells the group what he or she remembers. Other group members look back in the text to see if they have anything to add.

4. The *leader* asks each member of the group to share anything that was confusing (marked ? or x).

5. The group should try to figure out what the confusing word, sentence, or idea means. If the group cannot make sense of the confusing word, sentence, or idea, the *flag flyer* should raise the flag.

6. Repeat Steps 1 through 5 for Section 2. This time the *interpreter* shares what he or she remembers.

7. Repeat Steps 1 through 5 for Section 3. This time the *flag flyer* shares what he or she remembers.

8. The group should work together to answer the Big Question. The *interpreter* will write the group's answer to turn in to the teacher.

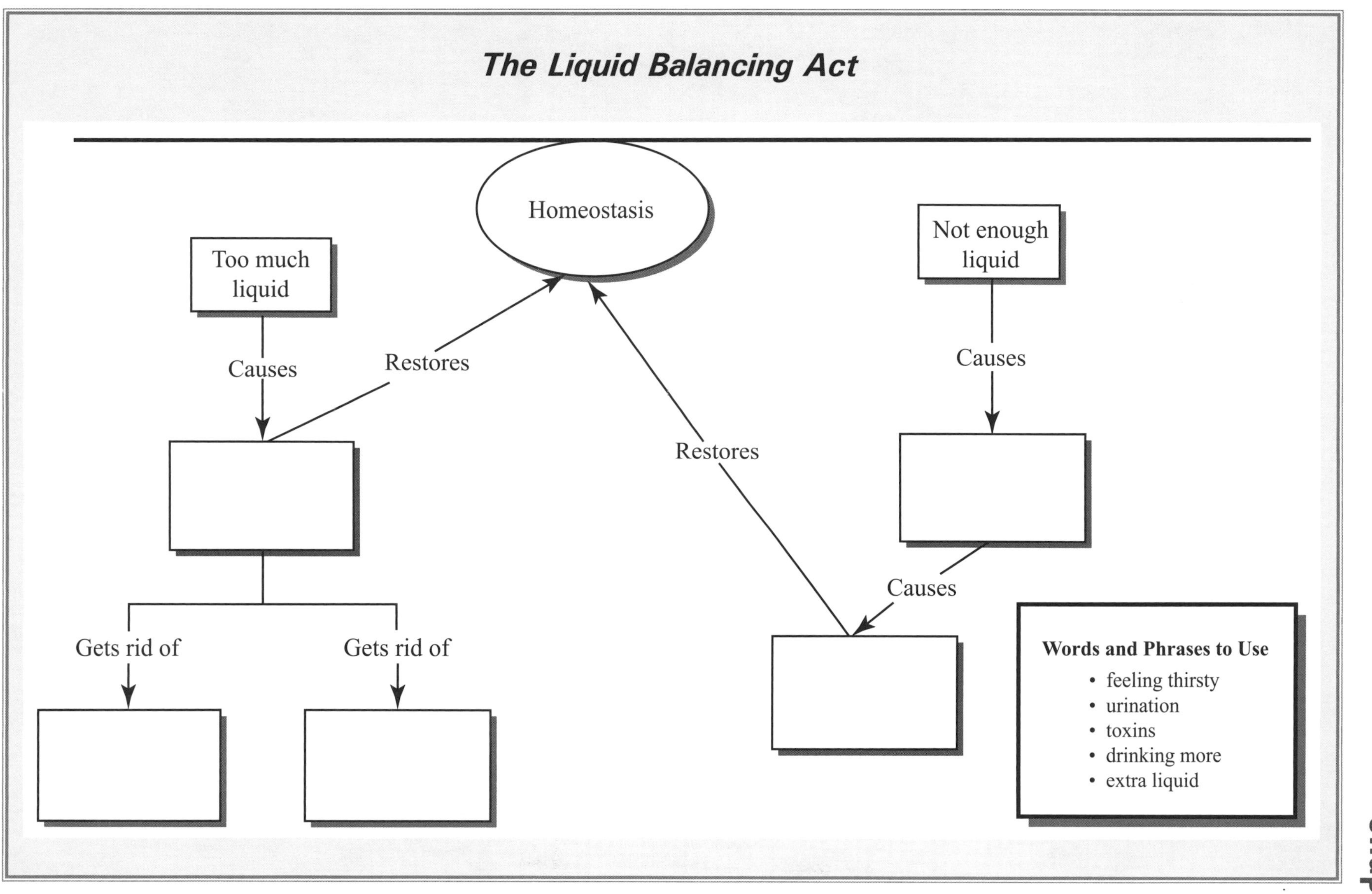
The Liquid Balancing Act
Homeostasis
Too much liquid
Causes
Restores
Gets rid of
Gets rid of
Not enough liquid
Causes
Restores
Causes
Words and Phrases to Use
• feeling thirsty
• urination
• toxins
• drinking more
• extra liquid

Chapter 11

A Crisis of Crabs

Topics

- Food chains and webs
- Biotic and abiotic factors
- Interconnections in ecology

NSES Content Standards (For Grades 5–8, Life Science)

- A population consists of all individuals of a species that occur together at a given place and time. All populations living together and the physical factors with which they interact compose an ecosystem.
- Populations of organisms can be categorized by the function they serve in an ecosystem. Plants and some microorganisms are producers—they make their own food. All animals, including humans, are consumers, which obtain food by eating other organisms. Decomposers, primarily bacteria and fungi, are consumers that use waste materials and dead organisms for food. Food webs identify the relationships among producers, consumers, and decomposers in an ecosystem.
- The number of organisms an ecosystem can support depends on the resources available and abiotic factors, such as quantity of light and water, range of temperatures, and soil composition. Given adequate biotic and abiotic resources, and no disease or predators, populations (including humans) increase at rapid rates. Lack of resources and other factors, such as predation and climate, limit the growth of populations in specific niches in the ecosystem. (NRC 1996, pp. 157–158)

Topic: Food Chains and Food Webs
Go to: *www.scilinks.org*
Code: LSB021

SAFETY ALERT!

- Keep clear of out-of-doors areas that may have been treated with pesticides, fungicides, and other hazardous chemicals.
- When working out-of-doors, students should use appropriate personal protective equipment (PPE), including safety glasses or goggles, gloves, close-toed shoes, hat, long-sleeve shirt and pants, sunglasses, and sun screen protection.
- Caution students relative to poisonous plants (ivy, sumac, etc.), insects (bees, wasps, ticks, mosquitoes, etc.), and hazardous debris (broken glass, other sharps, etc.).
- Teachers need to inform parents or guardians in writing of on-site or off-site field trips relative to potential health and safety precautions being taken.
- Teachers need to check with the school nurse relative to students' medical issues, such as allergies, asthma, and so on. Be prepared for medical emergencies.
- Teachers need to have a form of communication available, such as a cell phone or two-way radio, in case of emergencies.
- Wash hands with soap and water after doing out-of-doors activities.
- Be certain to contact the main office prior to bringing classes out of the building for science activities.

Reading Strategy

Pause, retell, and compare

Background

It is easy for ecology to degenerate into lists of vocabulary words and isolated concepts, leaving students without an appreciation for the complexity of ecological systems. This chapter is designed to help students think about the connections between all of the players in both a familiar and a less-familiar ecosystem.

This chapter would work well as a concluding activity for an ecology unit or following lessons on food chains, geological cycles (such as water, carbon, and nitrogen), and habitats. It could also introduce an ecology unit, assuming students have had some exposure to ecological concepts in elementary school.

Materials

- Access to a familiar ecosystem, such as the school yard
- Construction paper
- Strips of paper, approximately 3 in. long and wide enough for writing
- Glue
- Markers

Student Pages

- "A Crisis of Crabs"
- The Life of a Blue Crab

Exploration/Pre-Reading

In this exploration, students will look for connections in their school yard or other familiar environment. They will go outside and identify seven elements of that ecosystem, then think of ways that those elements affect one other.

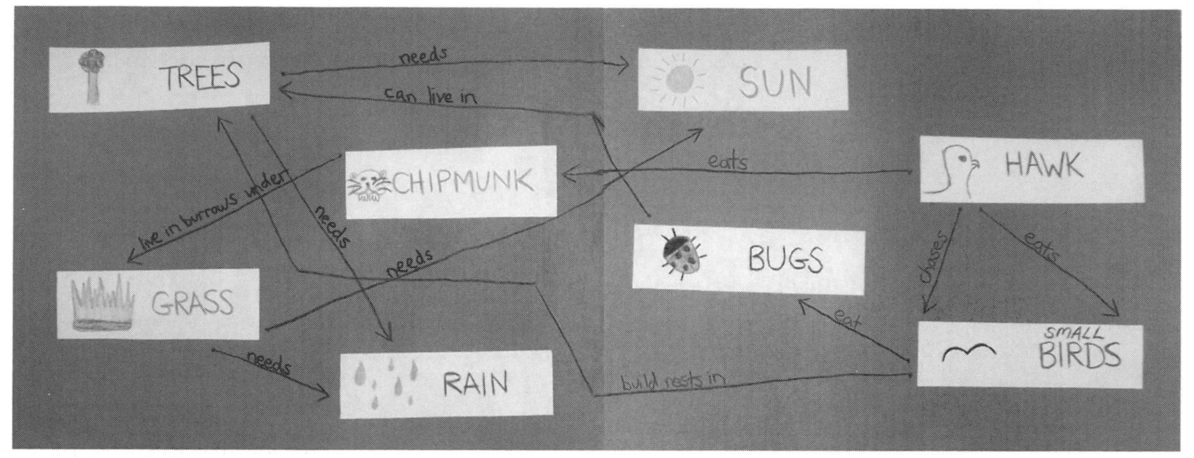

Source: Art by Brooke Robinson

Before class, locate an ecosystem for students to explore. An area that is at least partially "wild" (with minimal landscaping, for example) is preferable, but any basic school yard will do. If you teach at an urban school with no green space, consider a park in walking distance that you could visit with your students, a fish tank, or a terrarium. It is important that students actually explore the ecosystem in person. They will notice things that they might miss if they were asked to simply think about an ecosystem.

Topic: How Do Ecosystems Change Naturally?
Go to: *www.scilinks.org*
Code: LSB022

Give students seven to ten minutes to explore the ecosystem and list as many elements as they can. Stress that they should include nonliving things, such as the Sun and rain, as well as living things. They can list anything they can see, hear, or find evidence for. For example, if they spot an empty bird's nest, they can include a bird on their list.

When students return to class, they should select seven items from their list, including at least two nonliving things, and write each one on a slip of paper. Tell students that they are going to create a diagram that shows as many connections as they can think of between those parts of the environment. They should arrange the slips of paper in whatever system seems logical to them and then glue the slips to construction paper. Then they should draw arrows between connected elements and describe the connection. You may need to show them an example, such as a line connecting *Sun* and *grass* with the label *gives energy to grow*.

Conclude with this follow-up question:

Select one item from your diagram. Imagine that it was no longer present in this ecosystem. List as many changes as you can think of that would result from the item's absence.

Introduce the Reading. Tell students that they are going to read a passage that describes another ecosystem. This one is in the ocean, and the parts are connected in some complicated ways. Point out that the passage is written in first person—the author is reflecting on her own experiences exploring the ecosystem.

Reading Strategy: Pause, Retell, and Compare

Pause, retell, and compare was first introduced in Chapter 10. If you did not use Chapter 10 with your students, look at the strategy introduction described in that chapter. Put the first paragraph of "A Crisis of Crabs" on the board and model for students how to use pause, retell, and compare for that paragraph.

If you have already introduced pause, retell, and compare, ask a student to remind the class how the strategy works. Then place the first paragraph of "A Crisis of Crabs" on the board and have another student model using the strategy. If students are working in reading groups, you may need to provide them another copy of the modified procedure so they will remember how to use pause, retell, and compare in their groups.

Journal Questions

Have you tried using pause, retell, and compare when you read on your own? Think of some of the things you might have to read in the next week. Which readings would be easier to understand if you used this strategy?

> ***TEACHING NOTE***
> This graphic organizer is more complex than those in other chapters. If your students seem overwhelmed, consider displaying a copy and beginning the activity as a class. After a few blanks have been filled in, students will generally be comfortable completing it on their own.

Application/Post-Reading

- Graphic Organizer: The Life of a Blue Crab
- Pulling It Together in Writing: Make sure that students complete the graphic organizer that goes with "A Crisis of Crabs," either individually or in groups. Students will need access to this concept map, as well as the diagram they made during the exploration. Have them review both diagrams and then answer the following questions:

Study the diagrams that you have of your school yard ecosystem and the ecosystem of the blue crab. What similarities do you see? What differences are there? Which similarities would you expect to see in many other ecosystems?

- Pulling-It-Together Focus Point: All ecosystems are interconnected in many ways (e.g., through food chains and habitats and through relying on abiotic factors for survival). Changes to one element can change many others.

References

Lee, R., and M. Frischer. 2004. The decline of the blue crab. *American Scientist* 92 (6): 548–553.

Messick, G., and J. Shields. 2000. Epizootiology of the parasitic dinoflagellate *Hematodinium* sp. in the American blue crab *Callinectes sapidus*. *Diseases of Aquatic Organisms* 43 (2): 139–152.

National Resource Council (NRC). 1996. *National science education standards*. Washington, DC: National Academies Press.

Steele, P. 1979. A synopsis of the biology of the blue crab *Callinectus sapidus* Rathbun in Florida. Proceedings of the Blue Crab Colloquium, Biloxi, Mississipi.

REMEMBER YOUR CODES

! This is important.
✓ I knew that.
X This is different from what I thought.
? I don't understand.

A Crisis of Crabs

A food chain seems like a simple thing. A fish eats a plant. A crab eats the fish. A seagull eats the crab. Looking at a chain like that, it's easy to predict how changes will affect each part of the chain. If there are more plants, the fish will thrive. If there are too many seagulls, the crab population will drop. But when you throw in multiple life stages, a parasite, and some environmental factors, the picture quickly becomes more complicated.

I was on a boat off the coast of Georgia learning about just such a complicated situation—the sudden decrease in the blue crabs that support Georgia's crab fisheries. From 1999 through 2002, Georgia's blue crab harvest dropped to less than a quarter of its usual size. All of the state's crab-processing plants shut down, and crab fishermen who had been working off the Georgia coast for decades had to find other work.

Topic: Aquatic Ecosystems
Go to: *www.scilinks.org*
Code: LSB023

I leaned over a microscope on an inside deck and tried to hold my head steady against the sway of the boat. The thing I was looking at resembled a deformed chicken with a tail, only it was so tiny that it was barely visible with the naked eye. We had pulled it out of the water in a finely woven net that we towed behind the boat.

A Crab Matures

One of the scientists on board leaned over my shoulder. "That's a zoea you're looking

Life Cycle of a Blue Crab

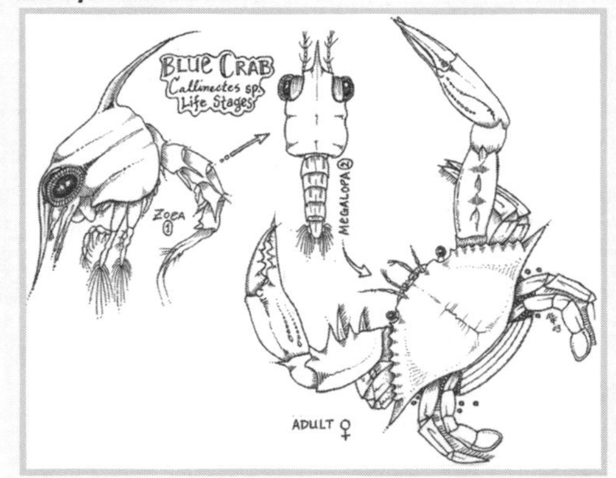

Source: Drawing by Michael G. Frick, research coordinator, Caretta Research Project.

at," he told me. "It's a blue crab just after it is born."

I was familiar with how a caterpillar turns into a butterfly. But I had no idea that other animals had similar changes. The blue crab, it turns out, comes out of its egg as this little zoea. It drifts in the currents out at sea, eating tiny sea creatures and hoping to escape the sea trout, croaker, and jellyfish that would love to gobble it up. If it survives, it will transform into its next stage after about a month. In its next stage, as a megalopa, it swims into the tidal currents and travels toward shore.

Into the Estuary

Our boat headed in, too. Another scientist pointed over the bow. "That's where we're going," he explained. "It's an estuary, the place where freshwater rivers meet the ocean. The freshwater dilutes the salt in the ocean water and provides a low-salt environment that supports all kinds of living things."

In the estuary, the megalopa transform into their more familiar crab shape. At first, the crab is only about 2 mm wide. It grows quickly, though, and within a year it will become an adult crab, as wide as the length of a pencil. Only one out of a million crab eggs survives to become an adult. The rest serve as important food sources for other species.

An Adult Blue Crab on the Sand

Source: Mary Hollinger, NODC biologist, NOAA. *www.photolib.noaa.gov/htmls/line0840.htm*

As an adult, the blue crab eats fish, worms, oysters, and other types of crabs. Blue crabs even eat other blue crabs. It turns out that this cannibalistic habit helped cause the decrease in crabs.

Even an adult crab, with its fierce claws, has predators. Adult crabs are eaten by large fish, seabirds, and raccoons, and they are parasitized by a tiny protist called *Hematodinium*.

A Crab Killer

"We think the *Hematodinium* are what's wiping out the crab population," a scientist told me. "They live inside the crabs and destroy their blue blood cells, the same blood cells that give the crabs their color."

"But the *Hematodinium* have been around for a while," he went on. "What we've been figuring out is why they are suddenly killing so many more crabs."

Why Now?

We pulled into the estuary and pulled up a water sample. We measured the salinity, or the amount of salt. Salinity had been one of the scientists' first clues. It measured 30 parts per thousand, almost as high as seawater.

"*Hematodinium* thrives in high-salt environments," the scientist explained. "It can't survive at all in freshwater. Usually, the estuary has relatively low salinity, but this year things have been different."

For the previous three years, Georgia had been in a severe drought. With low rainfall, the rivers that fed the estuary had been low, so less freshwater washed into the sea. Salinity in the estuary got higher and higher, providing a perfect environment for *Hematodinium* to spread.

One Life Stage of *Hematodinium*, Shown About 1,000 Times Bigger Than in Real Life

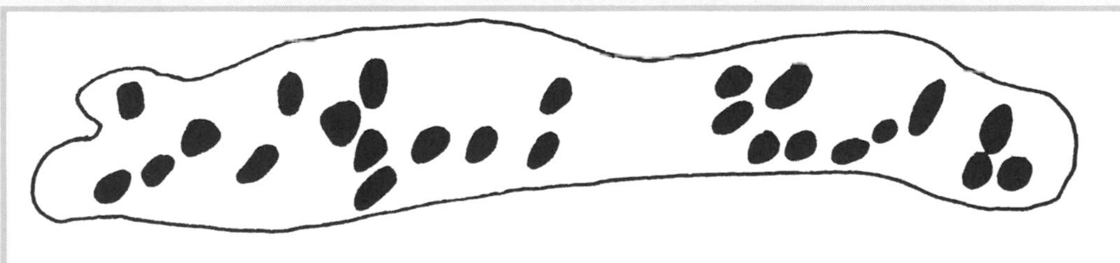

The scientist pointed to another clue, a temperature chart. The winter of 2001–2002 had been much warmer than usual. "The blue crab immune system is more effective in cold water," he explained. "Usually, the crabs can protect themselves in the winter. But this winter, *Hematodinium* raged all season long." As crabs got sick, other crabs ate them, spreading the infection rapidly through the population.

Ripples in the Food Chain

As our boat approached the dock, I asked how the blue crab disease was affecting the rest of the animals in this ocean food chain.

"We haven't studied that in depth yet," a scientist told me. "We've seen more porcelain crabs this year and a crab called the lesser blue crab. Blue crabs usually eat both of those kinds of crabs."

I pointed out that blue crabs spend a lot of their lives as prey, tiny zoea and megalopa that feed animals deep in the ocean. Were other animals going hungry?

"Hopefully, we won't have to find out," said another scientist. "Rainfall has been up this year, and the blue crabs are coming back."

I thanked the scientists for their time and got off the boat. As I stared back over the ocean, I had a new appreciation for complicated systems in nature, where low rain and a warm winter on land have the potential to ripple through the food chain for miles into the ocean.

THE BIG QUESTION

List three ways that the death of the blue crabs could affect other species in this ecosystem.

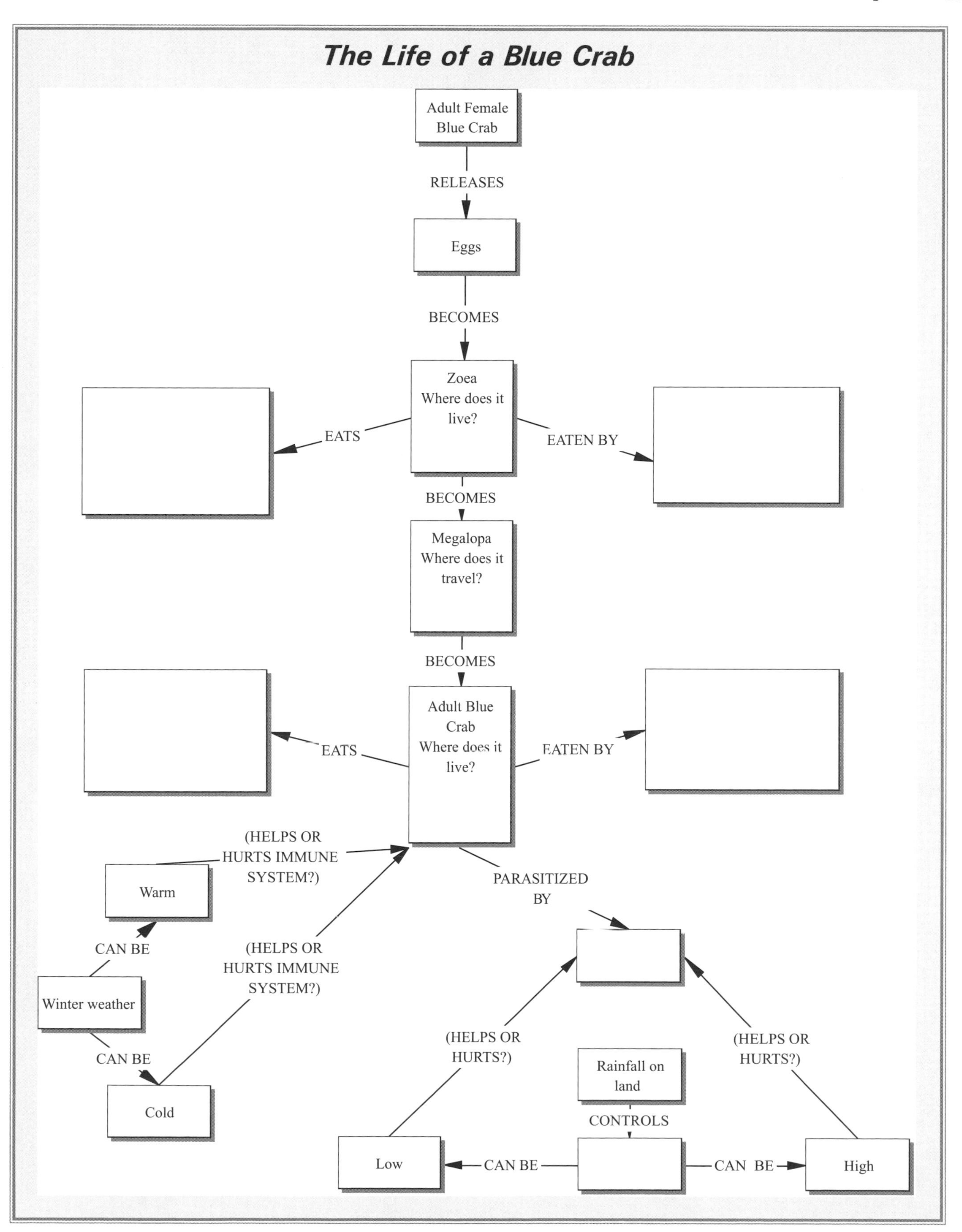
The Life of a Blue Crab
Adult Female Blue Crab
RELEASES
Eggs
BECOMES
Zoea Where does it live?
EATS
EATEN BY
BECOMES
Megalopa Where does it travel?
BECOMES
Adult Blue Crab Where does it live?
EATS
EATEN BY
(HELPS OR HURTS IMMUNE SYSTEM?)
Warm
CAN BE
Winter weather
CAN BE
Cold
(HELPS OR HURTS IMMUNE SYSTEM?)
PARASITIZED BY
(HELPS OR HURTS?)
(HELPS OR HURTS?)
Rainfall on land
CONTROLS
Low
CAN BE
CAN BE
High

Chapter 12

The Outsiders

Topics

- Classification
- Tentative nature of science
- Protists

NSES Content Standards (For Grades 5–8, Life Science and Science as Inquiry)

- Millions of species of animals, plants, and microorganisms are alive today. Although different species might look dissimilar, the unity among organisms becomes apparent from an analysis of internal structures, the similarity of their chemical processes, and the evidence of common ancestry.
- Scientific explanations emphasize evidence, have logically consistent arguments, and use scientific principles, models, and theories. The scientific community accepts and uses such explanations until displaced by better scientific ones. When such displacement occurs, science advances. (NRC 1996, p. 158, p. 148)

Topic: Levels of Classification
Go to: *www.scilinks.org*
Code: LSB024

Reading Strategy

Chunking

Background

Students often believe that classification is an innate system in nature, discovered whole and unchanged by scientists. In this chapter, students will get a brief introduction to how classification has changed over the years by studying the odd organisms in Kingdom Protista. Most students have had little, if any, exposure to protists prior to middle school. This chapter fits well after students have studied basic classification principles, including the levels of classification, and as an introduction to protists.

Materials

- Glass jar holding a culture of *Euglena*
- Welled slides
- Microscopes
- Pipettes
- Light source
- (Alternate exploration: a short video clip of *Euglena* moving)

SAFETY ALERT!

- Wash hands with soap and water upon completing this activity.
- Electrical light sources for microscopes need to be used only with a GFI-protected circuit.
- Never use your mouth to pipette liquids. Only use pipettes with bulbs attached.

Student Pages

- "The Outsiders"
- The Mysterious *Euglena*: Plant or Animal?
- Sorting Out the Living Things
- A Deadly Disease

Exploration/Pre-Reading

In this exploration, students will study *Euglena* and try to determine if they are plants or animals.

Before class, place the *Euglena* culture in front of the light source. Within a few minutes, the *Euglena* will have concentrated where the light hits the container.

Tell students that they will be looking at a new organism called a *Euglena*. Their job is to determine if it is a plant or an animal. Give students the handout The Mysterious *Euglena* and ask them to begin by observing the *Euglena* in the glass jar. What do they notice about where the *Euglenas* are located? Have them record their observation under Question 2 on the handout.

Show students how to use a pipette to collect a few *Euglena* from the cluster by the light source and place them in a welled slide. Instruct them to focus carefully so they do not immerse the microscope objective in the water. They will only need to use the low-power objective.

When the students have completed the handout, take a tally of those who think the *Euglena* should be a plant and those who think it should be an animal. Make a list of the evidence that students have gathered for both points of view. If students do not make the connection between the green color and the presence of chloroplast, ask them what other organisms are green and what causes that green color. At this point, some students may want to argue that it is both a plant and an animal or that it is neither. Allow those answers as well, but do not confirm the correct answer.

If you do not have the resources to allow students to observe live *Euglena*, a short film clip can be used as a substitute. Many short videos, taken through a microscope, are available on the internet, including a number of good options at the Euglenoid Project's website (*http://euglena.msu.edu/movies.shtml*). If your video includes sound, mute it so that students are not distracted by what the narrator may tell them about the organism. Explain that the *Euglena* has been videotaped through a microscope; its actual size is closer to the tip of a pencil. Have them complete the handout The Mysterious *Euglena* as described above.

Introduce the Reading. Tell students that the text they will read, "The Outsiders," will tell them how scientists classify *Euglena*. Students can read to find out if their classification agrees with the current scientific classification.

Reading Strategy: Chunking

This strategy is first introduced in Chapter 9. If you have not used Chapter 9 with your students, explain that many sentences in science writing have a lot of ideas crammed together. It would be difficult to try to understand all of the ideas at one time, but if students break the sentence down into chunks, they can think about each piece individually.

Place the following sentence on the board:

Euglena, with their ability to move and their food-making chloroplast, pose a dilemma for anyone who tries to classify all living things into just two categories: plant and animal.

If you have introduced chunking before, ask a volunteer to place slash marks (/) in the places where he or she might chunk the sentence, and work with the sentence in the chunks that the student selects. If you are introducing the strategy for the first time, place the slash marks yourself in the following locations:

Euglena, / with their ability to move and their food-making chloroplast, / pose a dilemma / for anyone / who tries to classify all living things into just two categories: / plant and animal.

Remind students that *Euglena* is the name of the organism that they were just studying. Then show that the phrase *with their ability to move and their food-making chloroplast* is an extra piece of information stuck into the middle of the sentence. Science writing frequently uses this kind of inserted phrase. Demonstrate that students can read the sentence without that phrase before going back to see what information the inserted phrase adds.

For the next phrase, ask a student to remind the class what a dilemma is. Then have students observe that the next phrase tells what sort of person is put into the dilemma (as they were, when they did the exploration activity).

Then reread the entire sentence and revisit the inserted phrase. Ask "What information does this phrase add?" and "What dilemma do *Euglena* pose?"

Remind students that chunking a sentence is like eating a pie. No one can put the whole pie in his or her mouth at one time; everyone eats it bite by bite. Like eating, some people will take bigger bites than others. That means that some people will need to break a sentence into more chunks than others, and that is okay. For this article, students can separate the chunks using slashes, like you did on the board. When they are reading something they aren't allowed to write on, they can chunk it in their head or cover up the parts of the sentence they aren't thinking about.

Journal Questions

Do you typically "chunk" sentences when you are reading? When is this strategy useful for you?

Application/Post-Reading

- Graphic Organizer: Sorting Out the Living Things
- Pulling It Together in Writing: Some of the most economically important protists are parasites. Provide students with the handout A Deadly Disease, from which they can learn more about the protist that causes malaria and demonstrate their knowledge of the characteristics of protists.
- Pulling-It-Together Focus Point: Plasmodium belongs in the protist kingdom because it has a single cell (so it is not an animal or fungus), moves and lacks chloroplasts (so it is not a plant), and has a nucleus (so it is not a bacteria).

References

Adl, S., B. Leander, A. Simpson, J. Archibald, O. R. Anderson, D. Bass, S. Bowser, G. Brugerolle, M. Farmer, S. Karpov, M. Kolisko, C. Lane, D. Lodge, D. Mann, R. Meisterfeld, L. Mendoza, Ø. Moestrop, S. Mozley-Standridge, A. Smirnov, and F. Spiegel. 2007. Diversity, nomenclature, and taxonomy of protists. *Systematic Biology* 56 (4): 684-689.

National Resource Council (NRC). 1996. *National science education standards*. Washington, DC: National Academies Press.

Scamardella, J. M. 1999. Not plants or animals: A brief history of the origin of kingdoms protozoa, protista and protoctista. *International Microbiology* 2 (4): 207–216.

Simonite, T. 2005. Protists push animals aside in rule revamp. *Nature* 438 (7064): 8–9.

REMEMBER YOUR CODES
- ! This is important.
- ✓ I knew that.
- X This is different from what I thought.
- ? I don't understand.

The Outsiders

A *Euglena* swims in a small pond, but even if you were swimming right beside it, you wouldn't see it. That's because the *Euglena* is so small that it can only be seen through a microscope. If you looked through a microscope, though, you would see one strange creature. *Euglenas* are green and contain chloroplasts. They can make their own food by photosynthesis, just like plants. But unlike plants, *Euglenas* can move. Thin, whiplike structures called *flagella* propel them through the water. And *Euglenas* have an eyespot that can tell light from dark, so they can swim to a place where they can absorb sunlight.

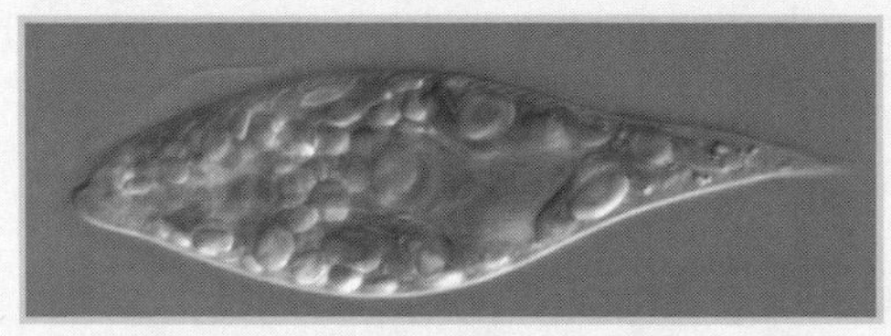

A *Euglena* swims in a drop of water.
Source: Photograph by Richard Triemer, Department of Plant Biology, Michigan State University

Scientists first discovered tiny one-celled creatures such as the *Euglena* when Antoni van Leeuwenhoek created a powerful microscope around 1700. He saw hundreds of tiny living things moving around in drops of water and bits of dirt. These new organisms included things we now know as bacteria, as well as thousands of other strange creatures. The organisms confused scientists because up until that point scientists had divided all living things into the plant and animal kingdoms. Where should they put the *Euglena*? It moved like an animal but made its own food like a plant!

Other creatures were equally confusing. Take the *Amoeba*, for example. *Amoebas* are made of just one cell that is so big that it can sometimes be seen, just barely, without a microscope. *Amoebas* don't have a mouth to eat their food. They don't even have a set shape. The cytoplasm inside the cell pushes on the cell membrane, creating long fingers that ooze across the bottom of lakes and ponds. Then the rest of the body catches up and the amoeba changes shape again. If the *Amoeba* comes across something it wants to eat, it simply surrounds its prey and digests it. Scientist didn't think that the *Amoeba* seemed like a plant or an animal.

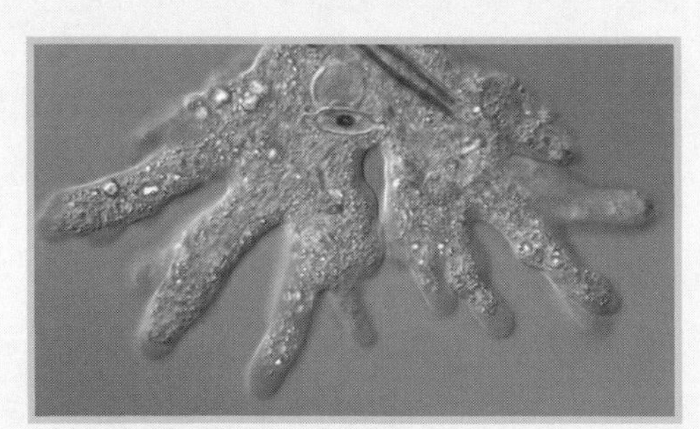

Amoebas constantly change shape.
Source: Photograph by Steve Durr

At first, biologists divided up the new one-celled creatures between the plant and animal kingdoms. *Amoebas* were classified as animals, bacteria were classified as plants, and the *Euglenas* were left homeless.

Then biologists tried again and created an extra kingdom to hold everything that didn't seem to be a plant or an animal.

However, in the late 1800s, biologists realized that bacteria were very different from amoebas and euglenas because bacterial cells do not have a nucleus. Plant and animal cells all have a nucleus. So do *Amoebas* and *Euglenas*. Bacteria were given their own kingdom because they could be defined as cells that do not have a nucleus. Eventually, the bacteria were divided between two kingdoms. However, it still wasn't clear what to do with the left-over creatures that had a nucleus but didn't seem to fit with plants or animals.

Another odd organism, the slime mold, takes "having a nucleus" to the extreme. It can have hundreds of nuclei in one big cell. Slime molds have a life cycle that is similar to a mushroom, and they decompose wood or other dead things like mushrooms do. Mushrooms belong in the fungus kingdom, but slime molds are different from mushrooms because they can move.

Slime molds crawl along on rotting logs or tree bark at an incredibly slow pace—1/25th of an inch per hour. At that rate, it takes eight days to move the length of a pencil! They even show some signs of intelligence—they can learn to complete a maze successfully to find food. A fungus or an animal? Scientists couldn't decide.

For now, all of the strange creatures that are not bacteria, animals, plants, or fungi are classified in a kingdom all by themselves: Kingdom Protista. Kingdom

This slime mold lives on a dogwood tree.
Source: John H. Ghent, USDA Forest Service, *www.Bugwood.org. www.forestryimages.org/browse/detail.cfm?imgnum=0796080*

Protista is different from the other kingdoms because the common ancestor of the protists is not known. All of the plants in the plant kingdom are closely related through evolution. The same is true for all of the animals and all of the fungi. Scientists do not think that all of the protists are as closely related. Kingdom Protista is simply a kingdom for the left-over organisms. Most scientists have added another grouping, bigger than a kingdom, to show that protists are related to all of the other living things that have a nucleus. There are three of these groupings, called *domains*, and all living things with a nucleus fall into the Eukarya domain.

The debate isn't settled, though. Scientists still argue over how to classify the protists. Some want to divide the protist kingdom into eight new kingdoms. Others want to divide them into even more kingdoms than that! But no matter how classifications change in the future, outsiders such as the *Euglena*, *Amoeba*, and slime mold will continue to interest scientists for a long, long time.

THE BIG QUESTION

Scientists classify living things into kingdoms, phylums, classes, and so on. Do they ever change the way they classify things? What might cause them to change the way something is classified?

The Mysterious Euglena: *Plant or Animal?*

Today, you are a scientist with the job of determining if this new organism, called *Euglena*, is a plant or an animal.

1. Watch several *Euglenas* under the microscope. Draw one of them here. (Make your drawing large enough to fill this space. Include as many details as you can see.)

2. Use words to describe what you observed.
(Some questions to consider: Does the organism move? If so, how? What color is it? What might give it that color? What behavior do you see?)

3. How do you think it should be classified—as a plant or an animal?

4. What evidence supports your answer?

A Deadly Disease

Malaria is a serious illness that kills hundreds of thousands of people, mostly children, each year. It is caused by the organism *Plasmodium* and is carried from person to person by mosquitoes. When an infected mosquito bites a person, the *Plasmodium* is injected into that person's blood. Eventually, the organism moves into the person's red blood cells and reproduces. Each *Plasmodium* consists of one single cell with a nucleus. It changes form several times during its life. Some of those forms can move on their own and some cannot.

When a mosquito infected with *Plasmodium* bites a person, it passes on the infection.

Source: Centers for Disease Control, *http://phil.cdc.gov/phil/quicksearch.asp*

If you had to classify *Plasmodium*, into which kingdom would you put it?

What evidence supports your answer?

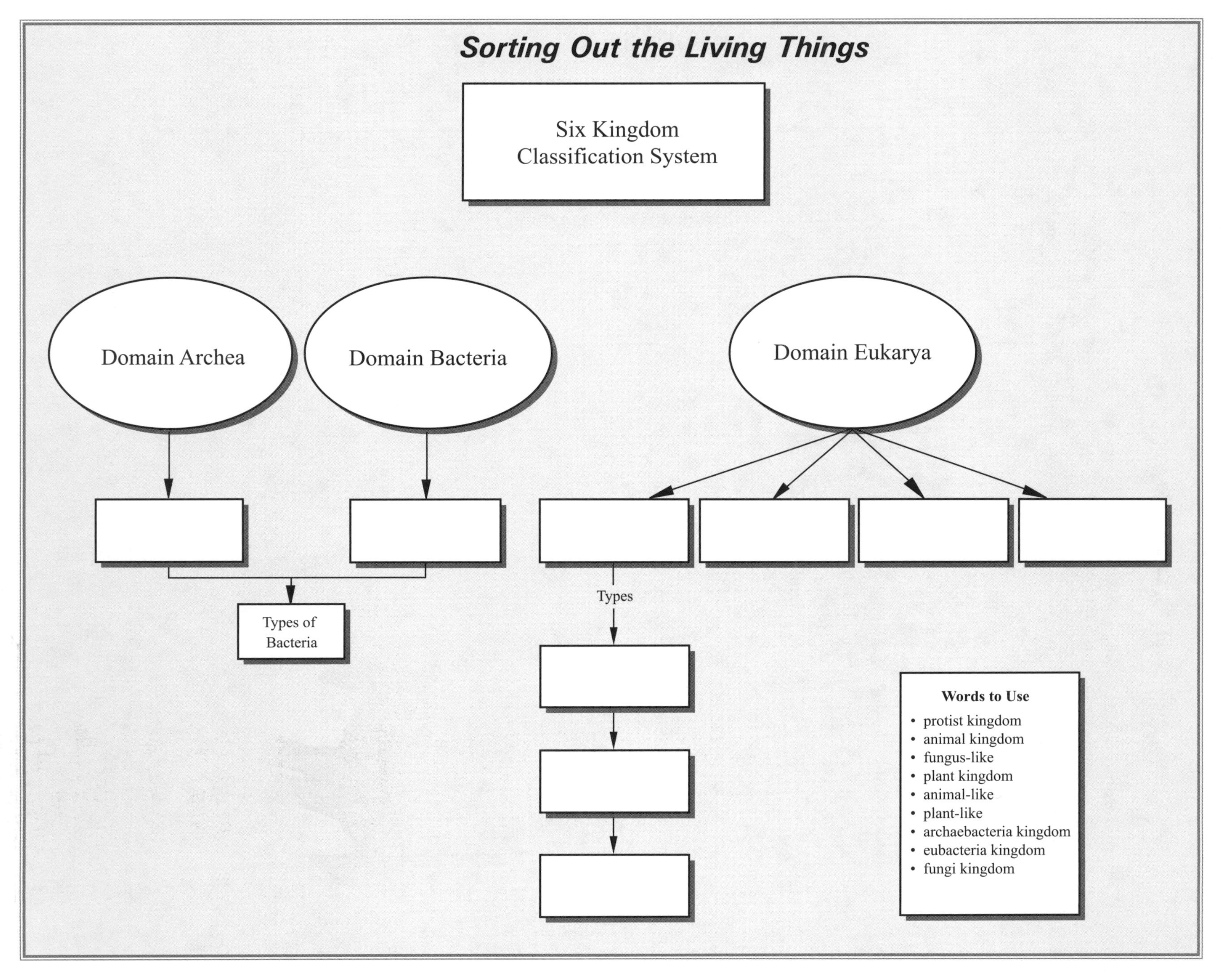
Sorting Out the Living Things
Six Kingdom Classification System
Domain Archea
Domain Bacteria
Domain Eukarya
Types of Bacteria
Types
Words to Use
• protist kingdom
• animal kingdom
• fungus-like
• plant kingdom
• animal-like
• plant-like
• archaebacteria kingdom
• eubacteria kingdom
• fungi kingdom

Chapter 13

Some Like It Hot

Topics

- Adaptation
- Natural selection
- Penguins

NSES Content Standards (For Grades 5–8, Life Science)

- Biological evolution accounts for the diversity of species developed through gradual processes over many generations. Species acquire many of their unique characteristics through biological adaptation, which involves the selection of naturally occurring variations in populations. Biological adaptations include changes in structures, behaviors, or physiology that enhance survival and reproductive success in a particular environment.
- An organism's behavior evolves through adaptation to its environment. How a species moves, obtains food, reproduces, and responds to danger are based in the species' evolutionary history. (NRC 1996, pp. 157–158)

Topic: Animal Adaptations
Go to: *www.scilinks.org*
Code: LSB025

Reading Strategy

Identifying text signals for comparisons and contrasts

Background

The word *adaptation*, as it is used in everyday speech, refers to a choice that individuals make to adjust to a new environment. In biology, however, adaptation refers to changes in populations that result from natural selection. This chapter uses a simulation involving penguins to show how biological adaptation can take place.

Although most people think of penguins as cold-weather birds, there are actually 17 species of penguins spread across the southern hemisphere. One species, the Galápagos penguin, actually lives in the tropics. Penguins provide an excellent example of how a body can adapt to both cold and hot environments.

Penguins probably evolved in New Zealand around 60 million years ago, before the supercontinent of Gondwana had separated, leading to the creation of Antarctica. The climate in the region at that time was fairly temperate. Penguins spread north and south from New Zealand over the next 40 million years, resulting in several new species, including the well-known Emperor penguins in Antarctica and the less-well-known Galápagos penguins in the tropics.

Materials

- 1 "Original Penguin" for each student—half get small penguins and half get large penguins
- Colored pencils or crayons to decorate the penguins
- 2 pages of penguin offspring cards for each group of 4 to 6 students (can be used with multiple classes)
- 1 die per group (bags of inexpensive dice are available at party supply stores)

Student Pages

- "Some Like It Hot"
- Penguin Journeys
- Penguin Adaptations

Exploration/Pre-Reading

In this exploration, students will simulate a flock of penguins that has migrated to a new climate. They will see how helpful adaptations increase over several generations in the new location.

To begin, distribute an "original penguin" to all of your students, allowing them to select the size that they prefer. Say that their penguin is going on a journey and they will need to prepare their penguin. They can either give them extra feathers by drawing feathers on their feet and beak, or give them bare skin on their feet and beak by coloring those areas yellow. The catch is, they don't know where their penguins are going, so they will just have to decide if they should have extra feathers or not. When students are finished preparing their penguins, the class should have a mix of large penguins with extra feathers, large penguins without extra feathers, small penguins with extra feathers, and small penguins without extra feathers. Place students into groups of between four and six students, and try to have a variety of penguin body styles in each group.

For the first simulation, tell students that their flock of penguins was caught in a storm and blown toward the equator. The penguins have ended up on a very warm island with no way to get home. Ask "Is a large penguin or small penguin better suited to live in a warm place?" It may surprise students to learn that the small penguin is less likely to overheat. A small penguin has a greater surface area to volume ratio, which allows heat to escape more easily. A large body size conserves heat. They can see this in other animals: Polar bears are the largest bears, arctic hares are larger than their desert counterparts, and large people tend to be warmer.

Then ask if a penguin with or without extra feathers would survive more easily in the warm environment. It should be easier for students to see that extra feathers are insulating, like a coat, and therefore less suitable for warm weather. At this point, students will be ready to begin the simulation.

Give students the Penguin Journeys handout and lead them through the first simulation as a class. Class data can be easily collected for each round by having students raise their hands to show what type of penguin they are.

For the second simulation, have all students revert to their original penguins. Explain to students that the penguins' colony became overcrowded, so they traveled toward the South Pole to find more room. Ask which body type is best suited to this new cold environment. Students should be able to predict that the big penguins with extra feathers will do

> ***TEACHING NOTE***
> This activity relies on probability, so there is always the chance that an unlikely penguin will prevail. This is most likely in small classes. You can lessen the odds by assigning the initial penguin body types so that they are all fairly represented. If you still generate unexpected results, you can repeat the simulation.

best here. Make sure that students notice that the numbers they must roll on the dice to survive have changed for this simulation.

In most cases, the trend will be clear just by looking at the raw numbers in the data table. However, you may wish to have students make a bar graph showing the change in population for each type of penguin in each environment before having them answer the reflection questions.

Introduce the Reading. Remind students of the simulation and what they learned. Tell students that the text they will read will give them more information on how penguins have changed to live in a variety of habitats.

Reading Strategy: Text Signals—Compare and Contrast

Begin by displaying the following excerpt from the reading so that students can see it:

> *Most birds are built for flying, with lightweight, hollow wing bones. Penguins, however, are built for swimming and diving.*

Remind students that certain words are signals for what the text is about to say. (Text signals are first introduced in Chapter 7 with words that signal examples and lists. If you did not use Chapter 7 with your class, you may need to provide more background on text signals than is given here.)

Underline the word *however*, and explain that *however* is a text signal. It tells you that the new information is going to contrast, or be different from, the earlier information. Ask students what contrast is being made, and show them the similarity in the phrases *Most birds are built for flying* and *Penguins are built for swimming and diving*. Ask students to predict what information might be in the next sentence. You may need to point out that the first sentence involves *wing bones* to get students to predict that the next sentence will talk about wing bones on penguins. After you get students' predictions, show students the next sentence from the text: *They have dense, heavy bones that make their wings work like flippers.*

Next, ask the class to generate a list of other words and phrases that can indicate a contrast. Add any words the class does not list. Common text signals for contrasting include *however*, *in contrast*, *on the other hand*, *conversely*, and *whereas*. Words that sometimes indicate a contrast include *but*, *yet*, and *while*.

Then ask students to generate a list of words and phrases that indicate that something will be compared to something similar. Common text signals include *in the same way*, *just like*, *just as*, *likewise*, and *also*.

Introduce students to the reading selection "Some Like It Hot." Before they start working in their reading groups, have them scan for text signals that indicate comparisons and contrasts and underline any that they find. In this particular selection, most of the text signals will be related to contrasts. Remind students to look for the contrast or comparison that goes with these words as they read.

Journal Questions

Imagine that a friend asks you what a text signal is. What would you tell your friend? Would you recommend that your friend look for signal words? Why or why not?

Application/Post-Reading

- Graphic Organizer: Penguin Adaptations
- Pulling It Together in Writing: Foxes provide another good example of similar animals adapted for different environments. Locate photographs of arctic and fennec foxes. Many photographs are available online; for best results, look for one that shows an arctic fox in winter when its coat matches the snow. Likewise, the fennec fox is roughly the color of the desert sand where it lives. Be sure to tell students that the fennec fox is the smallest of all foxes.

After students have looked at the photographs, have them answer these questions: What physical adaptations do these foxes have that make them suited to their environment? What behavioral adaptations can you think of that might be useful for these foxes?

- Pulling-It-Together Focus Point: Both foxes are colored to blend in with their environments to escape predators. The fennec fox is small and has large ears, so it loses heat easily. Behavioral adaptations could include any way that the arctic fox could conserve heat and the fennec fox could avoid it.

References

Chester, J. 1996. *The world of penguins*. San Francisco, CA: Sierra Club Books.
Davis, L. S. 2007. *Smithsonian Q & A: Penguins*. New York: HarperCollins.
Lynch, W. 2007. *Penguins of the world*. Buffalo, NY: Firefly Books.
National Resource Council (NRC). 1996. *National science education standards*. Washington, DC: National Academies Press.

REMEMBER YOUR CODES

! This is important.
✓ I knew that.
X This is different from what I thought.
? I don't understand.

Some Like It Hot

A flock of penguins gathers on the shore. They waddle clumsily across the rocks and then dive, one by one, into the refreshing ocean water. The cool water allows the penguins to escape the 100 degrees Fahrenheit heat on land.

What are penguins doing in a place with temperatures of 100 degrees? People typically picture penguins clustered in the ice and snow of Antarctica. There certainly are penguins in Antarctica, but there are also penguins up and down the coasts of Africa and South America. Penguins are even found in the tropics on the Galápagos Islands.

The Basic Penguin Plan

All penguins share basic features that make them penguins. They are birds, with two legs and two wings. Most birds are built for flying, with lightweight, hollow wing bones. Penguins, however, are built for swimming and diving. They have dense, heavy bones that make their wings work like flippers. When a penguin swims, it looks like it is flying under water.

All penguins—even those that live in the tropics—have thick, waterproof feathers that keep them warm and dry in the water. These feathers are organized in a pattern called *countershading*. When they swim, the black feathers on their backs help them hide from any predators looking down from above at the dark ocean water. Similarly, the white feathers on their bellies blend into the light of the sky if they are viewed from below. However, penguins from Antarctica to the Galápagos Islands have modifications of this body plan that fit the unique challenges of their own environments.

Emperor Penguins on the Ice

Source: Michael Van Woert, NOAA NESDIS, ORA. *www.photolib.noaa.gov/htmls/corp2395.htm*

Physical Differences

Emperor penguins live in the coldest environment. They travel across Antarctica to lay their eggs in the dead of winter, braving temperatures as low as −76 degrees Fahrenheit. They hunt squid, fish, and shrimplike animals called *krill* in water that is close to or below freezing. Their bodies are built for the cold. It's no accident that they are the largest penguins, with males weighing in at almost 85 pounds. Big bodies provide warmth. Feet and beaks, on the other hand, allow heat to escape. Compared to their body size, the feet and beak of an emperor penguin are the smallest of all penguins. Their feet and legs are covered with insulating feathers, and only their toes are bare for gripping the slick ice.

In contrast, it gets so hot on the Galápagos Islands that Galápagos penguin eggs can cook just by sitting in the sun. Therefore, Galápagos penguins have small bodies with big feet and big flippers. They have bare skin on their feet, on the underside of their flippers, and around their beaks. When they get hot, their blood concentrates in these areas to allow heat to escape.

Amazing Adaptations

Differences between animals that make them well suited to their environment, such as the body sizes of emperor and Galápagos penguins, are called *adaptations*. Adaptations are often physical characteristics; however, they can also be behavioral. Both emperor and Galápagos penguins have developed behavioral adaptations that help them survive extreme temperatures.

Behavioral Differences

Emperor penguin fathers incubate their eggs on the icy plains of Antarctica, where winds whip around them at up to 125 miles per hour. To stay warm, they hold their eggs on their feet and huddle together in large groups. When it storms, the group continually rearranges so that each dad gets a turn in the middle of the group, where it is warmest. They continue shuffling and rearranging for four long winter months until the mother penguins return.

Galápagos penguins, on the other hand, search out shade to incubate their eggs. They nest under rocks or bushes and in burrows to keep their eggs out of the Sun. They spend the hottest part of the day in the water, which is much cooler than land. When they return to shore, they hold their flippers out from their side to shade their feet and allow wind to circulate across the bare skin under their wings. They can't sweat, so they open their mouths and pant to create one more way for heat to escape.

Galápagos Penguin on a Rocky Island

Source:www.flickr.com/photos/putneymark/1351678861/sizes/l

Adaptations Develop Over Time

With unique adaptations, the basic penguin body plan works in both the icy Antarctic and the sweltering tropics. These adaptations developed over millions of years through natural selection. In each generation, those penguins that were best suited to their environment survived to produce chicks. Gradually, only the penguins with the most suitable adaptations remained. Emperor and Galápagos penguin adaptations illustrate how small changes can allow an animal to spread across a wide variety of environments.

Topic: Behaviors and Adaptations
Go to: *www.scilinks.org*
Code: LSB026

THE BIG QUESTION

What is an adaptation? Give one example from the emperor penguin and one from the Galápagos penguin.

Penguin Journeys

Name ______________________________

Is your original penguin ☐ big or ☐ small?

Does it have extra feathers? ☐ yes or ☐ no

Simulation 1: Travel to a Tropical Island

Step 1. Take a census. Count the number of each type of penguin in your class. Record the number on your data table.

Step 2. Can you survive the summer? Each member of the group will roll the die. The die roll tells if you survive. If you live, leave your penguin face up on your desk. If you die, turn it face down.

- If you are big with extra feathers, you must roll a 1 to survive.
- If you are big with no extra feathers, you must roll a 1 or 2 to survive.
- If you are small with extra feathers, you must roll a 1 or 2 to survive.
- If you are small with no extra feathers, you must roll a 1, 2, 3, 4, or 5 to survive.

Step 3. Take a census. Count the number of each type of surviving penguin in your class. Record the number in your data table for Round 1.

Step 4. Reproduce. If your penguin died, you get to become an offspring (a penguin chick). Look around at the surviving penguins in your small group. Which type of penguin is most common? Take a "penguin offspring" card that matches that type of penguin.

If there is a tie for most common, each of the surviving penguins should roll the die. Any penguins at your table that did not survive the last round should become an offspring of the highest roller.

Step 5. Take a class census. Record the numbers in your data table for Round 2.

Step 6. Repeat Steps 4 and 5 to complete Round 3.

Data Chart

	Big; Extra Feathers	Big; No Extra Feathers	Small; Extra Feathers	Small; No Extra Feathers
Starting Census				
Round 1				
Round 2				
Round 3				

Simulation 2: Fleeing to a Frozen Land

For this journey, everyone starts again with their original penguin. The steps are the same, except this time penguins that are big with extra feathers have the advantage.

Step 1. Take a census. Count the number of each type of penguin in your class. Record the number on your data table.

Step 2. Can you survive the winter? Each member of the group will roll the die. The die roll tells if you survive. If you live, leave your penguin face up on your desk. If you die, turn it face down.

- If you are big with extra feathers, you must roll a 1, 2, 3, 4, or 5 to survive.
- If you are big with no extra feathers, you must roll a 1 or 2 to survive.
- If you are small with extra feathers, you must roll a 1 or 2 to survive.
- If you are small with no extra feathers, you must roll a 1 to survive.

Step 3. Take a census. Count the number of each type of surviving penguin in your class. Record the number in your data table for Round 1.

Step 4. Reproduce. If your penguin died, you get to become an offspring (a penguin chick). Look around the surviving penguins in your small group. Which type of penguin is most common? Take a "penguin offspring" card that matches that type of penguin.

If there is a tie for most common, each of the surviving penguins should roll the die. Any penguins at your table that did not survive the last round should become an offspring of the highest roller.

Step 5. Take a class census. Record the numbers in your data table for Round 2.

Step 6. Repeat Steps 4 and 5 to complete Round 3.

Data Chart

	Big; Extra Feathers	Big; No Extra Feathers	Small; Extra Feathers	Small; No Extra Feathers
Starting Census				
Round 1				
Round 2				
Round 3				

Reflection Questions

1. In the first simulation, which type of penguin survived best in the warm environment?
2. Look at your data chart. How many of the penguins that survived the best did you start with at the beginning of the game? How many did you end with after Round 3?
3. What caused the change in the number of penguins of the different types?
4. What changes in penguin body type did you see in the second simulation? Describe what caused those changes to occur.
5. Did the penguins in these simulations get to choose what changes would happen in the population? How do you know?

Penguin Offspring	Penguin Offspring	Penguin Offspring
Small Bare skin on feet and beak	Small Bare skin on feet and beak	Small Bare skin on feet and beak
Penguin Offspring Small Extra feathers	Penguin Offspring Small Extra feathers	Penguin Offspring Small Extra feathers
Penguin Offspring Big Extra feathers	Penguin Offspring Big Extra feathers	Penguin Offspring Big Extra feathers
Penguin Offspring Big Bare skin on feet and beak	Penguin Offspring Big Bare skin on feet and beak	Penguin Offspring Big Bare skin on feet and beak

Source: Claire Wheeler

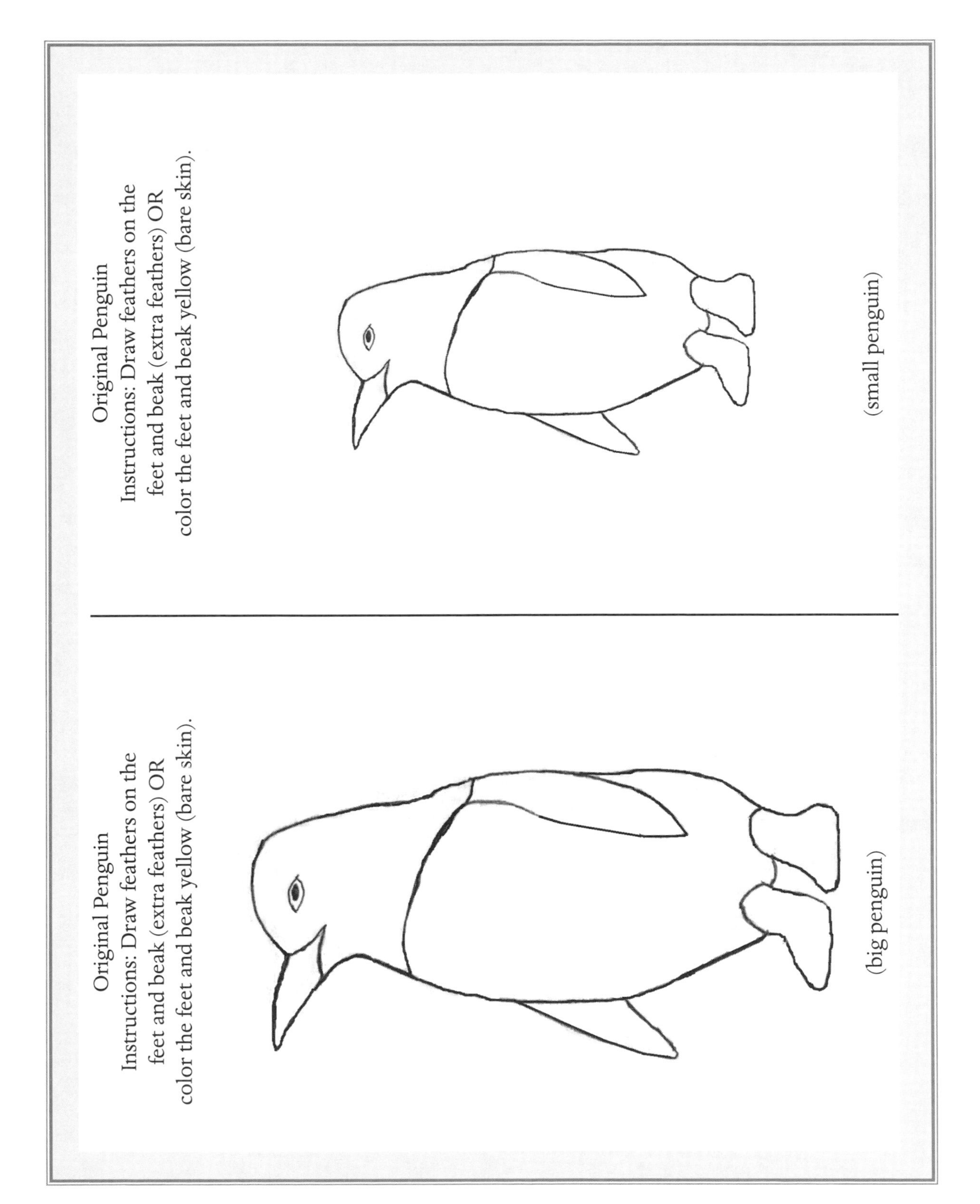
Original Penguin
Instructions: Draw feathers on the feet and beak (extra feathers) OR color the feet and beak yellow (bare skin).
(big penguin)
Original Penguin
Instructions: Draw feathers on the feet and beak (extra feathers) OR color the feet and beak yellow (bare skin).
(small penguin)

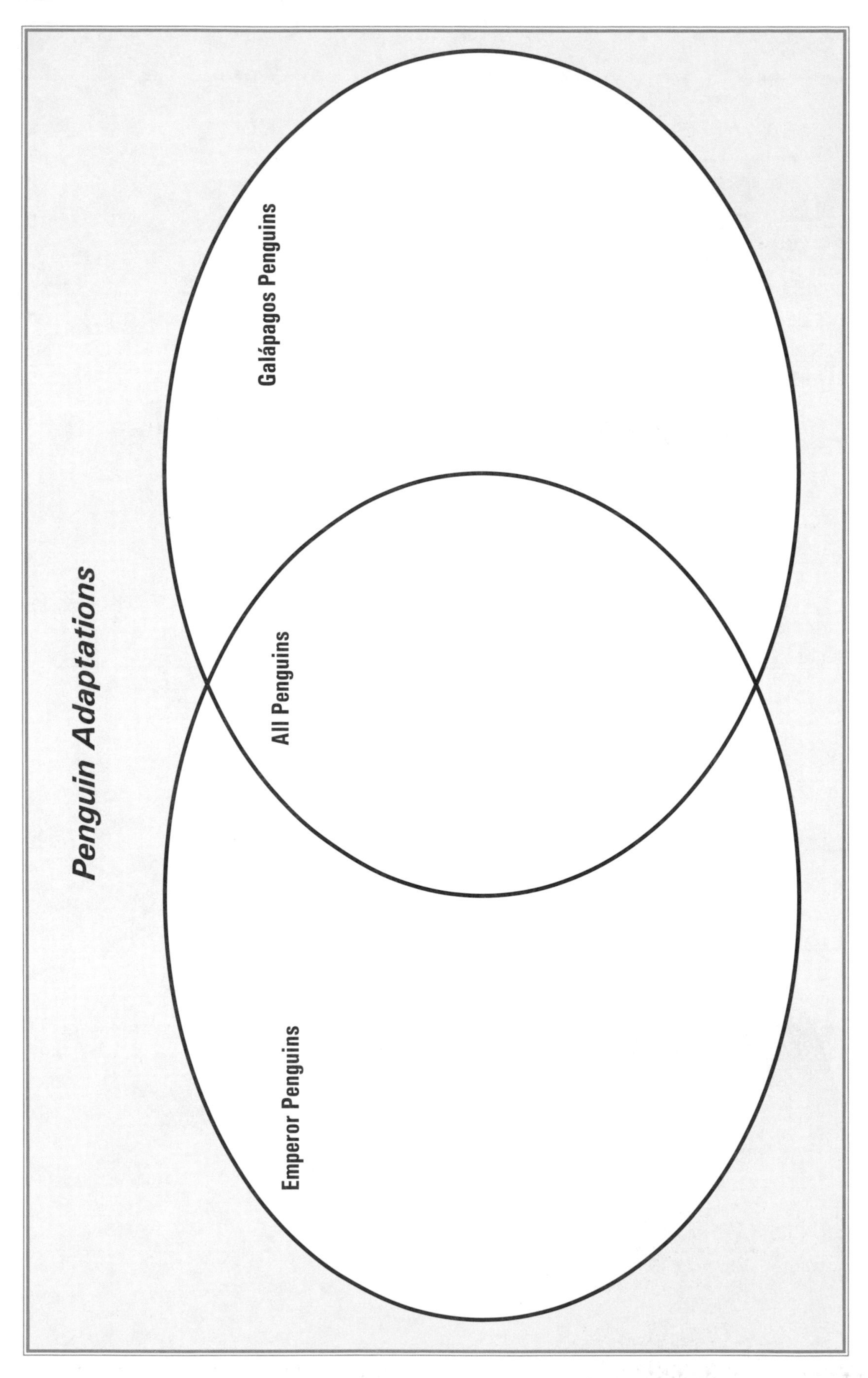
Penguin Adaptations
Galápagos Penguins
All Penguins
Emperor Penguins

Chapter 14

Bacteria: The Good, the Bad, and Getting Rid of the Ugly

Topics

- Bacteria structure
- Bacteria diversity
- Bacteria culturing

Topic: Bacteria
Go to: *www.scilinks.org*
Code: LSB027

NSES Content Standards (For Grades 5–8, Life Science)

- All organisms are composed of cells—the fundamental unit of life. Most organisms are single cells; other organisms, including humans, are multicellular.
- Cells carry on the many functions needed to sustain life. They grow and divide, thereby producing more cells. This requires that they take in nutrients, which they use to provide energy for the work that cells do and to make the materials that a cell or an organism needs.
- Disease is a breakdown in structures or functions of an organism. Some diseases are the result of intrinsic failures of the system. Others are the result of damage by infection by other organisms.
- Millions of species of animals, plants, and microorganisms are alive today. Although different species might look dissimilar, the unity among organisms becomes apparent from an analysis of internal structures, the similarity of their chemical processes, and the evidence of common ancestry. (NRC 1996, pp. 156–158)

Reading Strategy

Using context clues to find the meaning of new words

Background

Most students think about bacteria primarily in the context of disease. However, bacteria are the most numerous organisms on Earth, and only a fraction of them cause disease.

The exploration in this chapter—growing bacteria from the environment—is always popular with students. The reading reinforces their experience culturing bacteria, introduces the cell organization of bacteria, and clarifies that bacteria can be both helpful and harmful to humans.

> ***TEACHING NOTE***
> Local hospitals and laboratories are often willing to donate expired agar plates to schools. Expired plates are generally fine for this lab. Likewise, if you cannot locate sterile cotton swabs, you can substitute nonsterile swabs. Results will be less accurate but sufficient for the purposes of this lab.

> ***SAFETY ALERT!***
> Check with your school and district regarding policies on activities that involve culturing bacteria.

Materials

- 1 nutrient agar plate for each group of 2 to 4 students (available from scientific supply companies)
- Sterile cotton swabs (available in the first-aid section of a drugstore)
- 2 or 3 measurement grids (create by photocopying graph paper onto overhead transparencies)
- Clear tape or paraffin sealing wax
- Surgical gloves
- Ziploc bags

Student Pages

- "Bacteria: The Good, the Bad, and Getting Rid of the Ugly"
- Where Are the Most Bacteria in the Classroom?
- Bacteria, Bacteria, Bacteria

Exploration/Pre-Reading

In this exploration, students will design an experiment to test for the presence of bacteria in their classroom or school. Each group will identify three places they would like to test for bacteria and compare them to a control. Students will use cotton swabs to collect samples, plate the samples, and observe bacterial growth over the next few days.

Before class, draw lines with a permanent marker on the underside of each agar plate to divide the plate into four sections. Number the sections of the plate 1 through 4. One section will serve as a control, and the others will be used for the three places that students sample. Set out a bucket and label it *biological waste* so that students have a place to dispose of their used swabs on Day 1 and their used plates after the experiment is complete.

Figure 14.1. Agar Plate

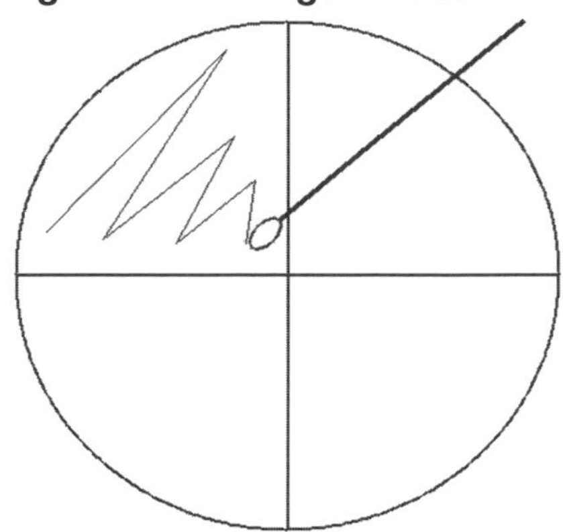

Gently swab the agar plate in a zig-zag motion.

Begin by having the class generate a list of places that they would expect to find bacteria. When designing their experiments, groups can use items off this list or come up with new ideas. Show students an agar plate and demonstrate how to swab a surface to collect bacteria and then gently wipe their cotton swab in a zig-zag motion on their plate (as seen in Figure 14.1). Alert students to common mistakes, which include swiping the outside of the plate instead of the agar, digging the swab into the agar, and placing the sample in the wrong section of the plate.

Hand out Where Are the Most Bacteria in the Classroom? Have each group decide on their three test locations and come up with a control. Most groups will choose to leave the control section blank, but a few will recognize the value of rubbing a clean swab in that section. Either control is acceptable. Likewise, some groups will be more insightful than others in thinking about the variables that they need to control. One important aspect they should consider is how they collect their samples (for example, number and length of swipes).

SAFETY ALERT!
Do not allow students to culture their body parts or fluids. This will lessen the chance that they will culture a disease-causing organism.

When the samples have been collected, use clear tape or paraffin sealing wax to secure the lids. If your school has an incubator, you can incubate the plates on low heat. Otherwise, keep the plates in a warm, dry location in your classroom. Colonies should grow in 24 to 48 hours. Be sure that students do not open the plates once they have been incubated.

SAFETY ALERT!
Because students could potentially culture pathogens, both plates and swabs should be treated as biological hazards. Make sure that students keep the plates closed and sealed after they have been cultured, and sterilize the plates and swabs after the experiment. If you have access to an autoclave, use it to sterilize the plates before disposing of them. A local hospital or laboratory may be willing to do this for you. Otherwise, soak the plates and swabs in bleach before throwing them away.

Tell students that the results for this experiment will give them an idea of the differences in bacteria concentration around the classroom. They won't be able to get an accurate count of the actual number of bacteria. Microbiologists have methods to determine accurate counts, but those methods require more materials.

There are two ways to calculate results. Some groups will have distinct colonies. They can simply count the number of colonies in each section. In other groups, the colonies will have run together. These groups can use a measurement grid to estimate the area that is covered in each section. Some plates may have large, hairy growths. Most of these are fungi rather than bacteria. Students may wish to account for the fungi separately. Drawing conclusions for this experiment is then rather straightforward.

Introduce the Reading. Tell students that this reading selection will tell them more about what they saw on their agar plates. In particular, they should be reading to find out exactly what colonies are.

Reading Strategy: Finding the Meaning of New Words

Note that this strategy is introduced in Chapter 5. If you have not used Chapter 5, introduce this strategy according to the instructions found there instead.

If your students have already been introduced to this strategy, begin by asking them if they can remember some of the ways that science books give the definitions of new words. List their answers on the board and add any that they may have forgotten (see Table 2.1 in Chapter 2).

Ask students if anyone has been successful finding the meaning of a new word lately. Have volunteers share what they were reading and what words they learned. Tell them that this reading also has a lot of new words and to remember to look before and after the new word in the sentence if they are having a hard time finding what it means.

At this time, you may also need to clarify that *bacteria* is plural, meaning many bacteria, while *bacterium* is singular, meaning just one.

Journal Questions

Do you have ever have trouble remembering the meaning of a new word after you read it? What could you do while you read to help you keep up with the meanings of new words?

Application/Post-Reading

- Graphic Organizer: Bacteria, Bacteria, Bacteria
- Pulling It Together in Writing: This exploration lends itself to sharing the results. Have your students write a letter to the principal or school newspaper describing their experiment and what they found. They should include relevant information from the reading. How concerned should the school be about their findings? Send the best letters to the principal or newspaper.
- Pulling-It-Together Focus Point: Responses will vary according to the results of their investigation, but (assuming investigation results support the statement) students should indicate that bacteria are common in the environment and not all bacteria are harmful.

References

Canby, T. Y. 1993. Bacteria: Teaching old bugs new tricks. *National Geographic* 184 (2): 36–61.

Darling, K. 2000. *There's a zoo on you*. Brookfield, CT: Millbrook Press.

Dyer, B. D. 2003. *A field guide to bacteria*. New York: Cornell University Press.

National Resource Council (NRC). 1996. *National science education standards*. Washington, DC: National Academies Press.

Salleh, A. 1999. Helpful bacteria help clean up toxic groundwater. Australian Broadcasting Corporation. *www.abc.net.au/science/news/stories/s54259.htm*

REMEMBER YOUR CODES
- ! This is important.
- ✓ I knew that.
- X This is different from what I thought.
- ? I don't understand.

Bacteria: The Good, the Bad, and Getting Rid of the Ugly

Bacteria are everywhere. A teaspoon of soil can have as many as a million bacteria. Your skin is home to 100 billion. And that doesn't even begin to cover what can be found in your mouth or intestines.

If you look closely at your skin, you won't see any bacteria. A single bacterium can only be seen with a high-powered microscope. Each bacterium is made of just one simple cell. Bacteria are *prokaryotes*, which means that they do not have a nucleus. Instead, they have a tangle of DNA that floats in the cytoplasm. They lack other organelles, too, such as Golgi bodies, endoplasmic reticulum, and vacuoles.

But as their numbers show, *simple* doesn't mean the bacteria aren't successful. Bacteria species survive in almost every imaginable place: our bodies, the ice in Antarctica, rotting wood, and even barrels of poisonous chemicals.

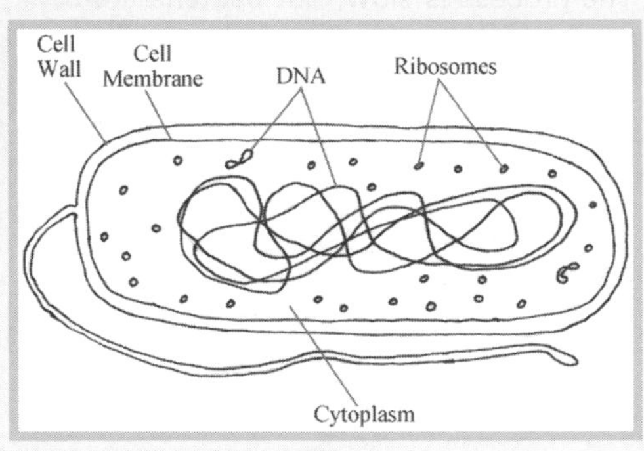

Bacterial cells are small and simple.

Not So Deadly

When most people think of bacteria, they think of disease. Bacteria cause a variety of human diseases, such as strep throat, whooping cough, and anthrax. But only about 4% of bacterial species are pathogens, or organisms that cause disease. Many bacteria are actually helpful to humans.

Your intestines are filled with bacteria that help with an important task—digesting your food. Our digestive system can't break down all of the food we eat, so the bacteria in our intestines dine on the leftovers. We provide food and housing for them, but they give us a lot in return. One of the most common bacteria in our intestines, *E. coli*, makes vitamin K, which helps our blood clot. It also makes folic acid, which helps prevent birth defects.

Another common intestinal bacteria, *Lactobacillus acidophilus*, protects our intestines from other bacteria and fungi that cause disease. *L. acidophilus* is found in yogurt. When you take antibiotics for an illness, those antibiotics also kill *L. acidophilus*. Eating yogurt can help replenish your supply.

Helpful Hunger

Wherever bacteria live, they have to find something to eat. Bacteria have evolved to eat just about everything. Humans are learning to take advantage of this to clean up pollution.

As you can imagine, gas stations have the ability to produce a lot of pollution. At a gas station, the storage tanks for gasoline are buried underground. After about fifteen years, the tanks get old and may begin to leak. Gasoline in the environment separates into several different chemicals, some of which cause cancer. Scientists who study bacteria, called *microbiologists*, have discovered bacteria that can digest the cancer-causing chemicals. These bacteria can be used to clean up leaked gasoline.

Microbiologists have found bacteria that eat other pollutants as well. There are bacteria that digest poisonous cyanide, pesticides, old paint, and even oil spills. The process is slow, but bacteria are being used to clean up polluted places all around the world.

Underground storage tanks may leak gasoline as they get older.

Source: Drawing by Claire Wheeler

Agar-den of Bacteria

Microbiologists have found ways to grow bacteria in the lab so they can study them. One common way to grow bacteria is to put them on an agar plate. An agar plate contains a jellylike substance (agar) and the nutrients that bacteria need in order to grow.

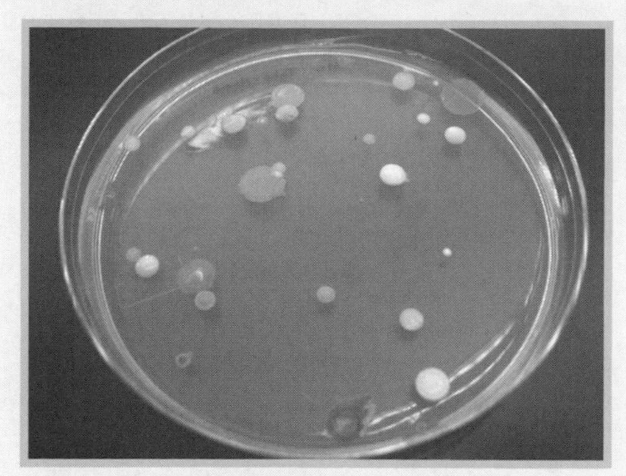

Bacteria grow in clumps called *colonies.*

Source: Minyoung Choi, *www.flickr.com/photos/mrmin123/1737892527*

Individual bacteria cannot be seen when they are rubbed onto an agar plate with a cotton swab. But the bacteria begin to reproduce very quickly. In just a day or two, one bacterium will divide enough times to make a clump of bacteria that can be seen by the naked eye. These clumps are called colonies. Each bacterial colony consists of thousands or even millions of bacteria that came from one original bacterium.

Not all types of bacteria grow well on agar, but most of the ones that live on and around humans grow nicely. You can use an agar plate to see some of the millions of bacteria in the world around you.

THE BIG QUESTION

Many people think all bacteria cause disease. Is this true? Write a paragraph that supports your answer.

Where Are the Most Bacteria in the Classroom?

Part 1: Background Knowledge

Bacteria can be grown on an agar plate. Agar is a gel, kind of like Jell-O, made from a substance in seaweed. Nutrients that the bacteria can use as food are added to the agar. You can use a cotton swab to collect bacteria that are too small to see and put them on an agar plate. In a few days, you will be able to see clumps of bacteria on the plate.

Part 2: Question

Where are the most bacteria in the classroom?

Part 3: Procedures

Your group will receive an agar plate divided into four sections. Three sections are for testing three different places. One section is for your control.

What three places will your group swab? (Be very specific!)

1.

2.

3.

4. Section 4 is your control.

A control is used to make sure that the bacteria you find growing in Sections 1, 2, and 3 really came from where you swabbed. For example, it's possible that bacteria might float in from the air when you open your plate. Without a control section, you'd never know.

What will you do to keep Section 4 as your control? ____________________

____

You also need to keep things the same for each place you swab to make sure that this is a fair test. What will you do to keep things fair?

Use a clean swab for each section.

____

____

Directions for collecting samples

1. Put on surgical gloves. Run the cotton swab over the area you are testing.
2. Open the plate and run the swab in a zig-zag motion over the top of the agar gel. Be sure to put it in the section with the same number as the place you are testing. Do not push the swab into the agar. You should not put a dent in it.
3. Close the plate. Do not walk around with the top off of your agar plate.
4. Put your used cotton swabs into the bucket labeled *biological waste*.
5. When all sections are done, tape the plate closed, seal it in a plastic bag, discard gloves into biological waste bucket, and wash your hands with soap and water.

Part 4: Data/Observations

DO NOT OPEN THE PLATES! Most of the bacteria will be harmless, but some could make you sick! After you are done with your plate, put it in the bucket labeled *biological waste*.

Draw what you see in your plate:

How many bacteria grew in each section?

Location	Number of Colonies or Grid Squares
1.	
2.	
3.	
Control	

Did anything grow in your control section?

If something grew in a group's control section, could they be sure the bacteria in the other sections came from where they swabbed? Why or why not?

Part 5: Claims and Evidence

Where did you find the most bacteria? Describe the evidence that supports your claim.

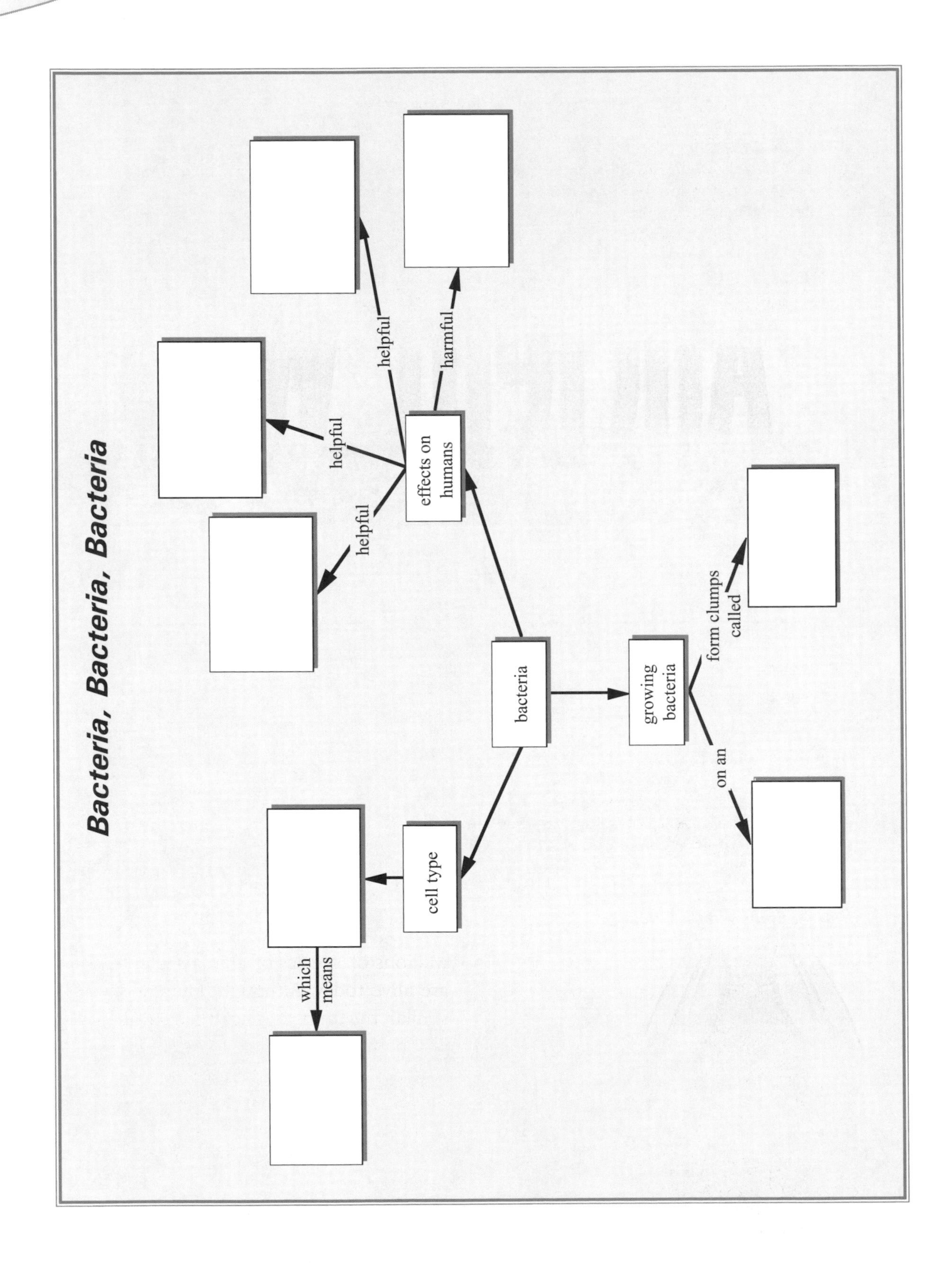

Bacteria, Bacteria, Bacteria
bacteria
effects on humans
helpful
helpful
helpful
harmful
cell type
which means
growing bacteria
form clumps called
on an

Chapter 15

Hunting the Ancient Whales

Topics

- Macroevolution
- Evidence for evolution
- Characteristics of mammals

NSES Content Standards (For Grades 5–8, Life Science)

- Millions of species of animals, plants, and microorganisms are alive today. Although different species might look dissimilar, the unity among organisms becomes apparent from an analysis of internal structures, the similarity of their chemical processes, and the evidence of common ancestry.
- Biological evolution accounts for the diversity of species developed through gradual processes over many generations. Species acquire many of their unique characteristics through biological adaptation, which involves the selection of naturally occurring variations in populations. Biological adaptations include changes in structures, behaviors, or physiology that enhance survival and reproductive success in a particular environment. (NRC 1996, p. 158)

Topic: History of Evolution
Go to: *www.scilinks.org*
Code: LSB028

Topic: Whales
Go to: *www.scilinks.org*
Code: LSB029

Reading Strategy

Recognizing and reading scientific names

Background

As mammals that live in the water, whales have been an enigma since the time of Darwin, who was repeatedly mocked for suggesting that perhaps they evolved from swimming bears. However, in the past 20 years, biologists and paleontologists have uncovered a remarkably complete story of whales' transition from land to water. The evolution of whales provides an excellent example of how biologists use a variety of types of evidence to understand evolution. This lesson introduces students to a small portion of the fossil, biochemical, and genetic information that is known about the evolution of whales.

Materials

- Copies of the "Whale Evolution Images"—1 per group of 2 or 3 students
- Photograph or picture of a modern whale
- Copy of "Ages of Some Whale Fossils" to display

Student Pages

- "Hunting the Ancient Whales"
- Evidence From the Ancient Whales
- Going Batty
- Relationships Between Some Whale Fossils
- Whale Evolution Images

Exploration/Pre-Reading

In this exploration, students will examine skeletons and reconstructions of some ancient and modern whales to look for a pattern of change. Before class, cut out the whale evolution images and place a complete set in an envelope for each group. Leave the skeletons and their reconstructions attached to each other.

Introduce the activity by showing the class a picture of a modern whale. Tell the class that a local museum has received pictures of whale fossils for a display, but they arrived scrambled. With their partners, students must take the scrambled pictures and try to arrange them in order from those that look most like land animals to those that look most like modern whales.

As they work, some students may ask about how the reconstructions have been derived from the skeletons. Let them know that some features of the reconstructions, such as the skin color, are artists' interpretations. However, skeletons give us more information about the complete animal than one might think. Scientists can compare the fossil skeletons to the skeletons of living things to see how the skeleton affects the shape of the body. Likewise, bones bear marks from soft tissue, such as muscle attachments, that can give us clues about the shape and size of those soft tissues.

After students have had a chance to arrange their fossils, display the page Relationships Between Some Whale Fossils, and let students compare their lineage with the lineage that dating methods used by scientists have shown. One important point to make for your students is that the sample fossils used here are not necessarily direct ancestors for modern whales. Instead, they represent branches of whale evolution that share common ancestry with modern whales.

Introduce the Reading. Tell students that today's reading will give them more information about the history and lives of some of these whales.

Reading Strategy: Reading Scientific Names

Begin by reviewing the concept of scientific names with students. Ask them "What are scientific names? Why do they exist?" Point out that scientific names are often long and can be confusing to a reader.

Ask students how they can know when they come across a long word if it is a scientific name. They may mention the context, that the name will be written in italics, or that the name will usually consist of two words (a genus and species). Point out that scientific names are usually written in italics, and that the genus is capitalized but the species is not. Sometimes only the genus is given (e.g., *Pakicetus*) or only the first letter of the genus with the full species (e.g., *E. coli*). Write the following names on the board, and tell students that the same animal might be called any of these names: *Pakicetus inachus*, *Pakicetus*, or *P. inachus*.

Have students brainstorm ways to handle reading the "Latinized" words. They may suggest sounding the word out, calling it by just the first letter, or giving it a nickname, such as the first syllable. Point out that the important thing is not that they pronounce it correctly but that they know they are reading a name.

Journal Questions

Which strategy did you use for reading the scientific names? Will you use the same strategy next time you come across scientific names? Why or why not?

Application/Post-Reading

- Graphic Organizer: Evidence From the Ancient Whales
- Pulling It Together in Writing: Bats pose a similar evolutionary problem to whales. They are also mammals, but they fly like birds. Bat evolution is not as well documented as whales because bat skeletons are tiny and do not fossilize well, but scientists are making some progress. Give students the handout Going Batty and ask them to think about how they would research the evolution of bats if they were scientists.
- Pulling-It-Together Focus Point: Bat fossils would tell about how bats have changed shape over time. Soil and chemicals surrounding the fossils could indicate the type of environment in which the fossil bats lived. DNA could help identify how bats are related to other animals. Bat gene patterns could indicate how bat wings developed.

References

Gingerich, P., M. ul-Haq, W. von Koenigswald, W. J. Sanders, B. H. Smith, I. S. Zalmout. 2009. New Protocetid Whale from the Middle Eocene of Pakistan: Birth on land, precocial development, and sexual dimorphism. *PLoS ONE* 4 (2): e4366.

Gingerich, P. 2004. Whale evolution. *The McGraw-Hill yearbook of science and technology,* 376–379. New York: McGraw-Hill.

National Resource Council (NRC). 1996. *National science education standards*. Washington, DC: National Academies Press.

Paine, S. 1995. *The world of the Arctic whales*. San Francisco: Sierra Club Books.

Thewissen, J. G. M., M. J. Cohn, L. S. Stevens, S. Bajpai, J. Heyning, W. E. Horton. 2006. Developmental basis for hind-limb loss in dolphins and the origin of the cetacean body plan. *Proceedings of the National Academy of Science* 103 (22): 8414–8418.

Thewissen, J. G. M., L. N. Cooper, J. C. George, and S. Bajpai. 2009. From land to water: Origins of whales, dolphins, and porpoises. *Evolution: Education and Outreach* 2 (2): 272–288.

Thewissen, J. G. M., and B. Sunil. 2001. Whale origins as a poster child for macroevolution. *BioScience* 51 (12): 1037–1049.

Thewissen, J. G. M., and E. M. Williams. 2002. The early radiations of Cetacea (Mammalia): Evolutionary pattern and developmental correlations. *Annual Review of Ecology and Systematics* 33 (1): 73–90.

Zimmer, C. 1998. *At the water's edge: Macroevolution and the transformation of life*. New York: Free Press.

REMEMBER YOUR CODES
! This is important.
✓ I knew that.
X This is different from what I thought.
? I don't understand.

Hunting the Ancient Whales

In the frozen arctic, a bowhead whale dives under a sheet of ice and swims as far as it can. After a few minutes, it rears its 60-ton body upward and smashes a hole in the ice. Bowhead whales have to come to the surface regularly because, like humans, whales breathe air.

A Bowhead Whale in the Arctic Ocean

Source: Photograph by Kristin Laidre

Humans and whales are both mammals. We both produce milk for our babies. We both have hair, although whales don't have very much. And some whales, like the bowhead, have even more in common with humans. They have small leg bones buried deep in their layers of fat. The whales don't need these bones. They are leftovers from millions of years ago, when whales walked on land. How did whales move from being land animals to sea creatures? Scientists have been gathering evidence from a variety of sources to try to answer that question.

The Trail Begins in Pakistan

The first set of clues came from the fossil record. The earliest fossils that scientists are sure belonged to a whale are from *Pakicetus inachus*. As soon as the first *Pakicetus* fossil was unearthed in Pakistan, scientists knew it was related to whales. All whales have a dense, thick casing for their ear bones that is unlike the ear casing of any other animal. At 50 million years old, *Pakicetus* is the oldest fossil that has this type of ear. However, *Pakicetus* had four legs. And at the end of each leg there were hooves like you would find on a pig, hippopotamus, or giraffe!

An Artist's Reconstruction of *Pakicetus*

Source: Illustrated by Carl Buell, and taken from *www.neoucom.edu/Depts/Anat/Thewissen/whale_origins/whales/Pakicetid.html*

Could whales be related to animals with hooves? Scientists used DNA to find out. Animals that are related have a lot of the same DNA. Scientists collected DNA from living whales and compared it to the DNA from other mammals. Sure enough, the closest match came from an animal with hooves—the hippopotamus.

To the Nearby Swamps

The next most recent fossils came from Pakistan and nearby India. One fossil in this group belonged to *Ambulocetus natans*, a 400-pound beast with sharp teeth and a long tail. Relative to its body size, *A. natans* had shorter legs than *Pakicetus*. It also had webbed feet. Its body shape suggested that it spent a lot of its life in water. But did it?

Scientists have many ways of finding out if fossils are from land or sea animals. They can study the rocks that surround the fossil. Rocks from land, deep parts of the ocean, and near the shore are all different. Scientists can look to see what else has been fossilized in those places. If the surrounding fossils are sea creatures, such as ancient marine clams or snails, then that adds more evidence that the area was once under water. They can even look to the chemical makeup of the teeth and bones. Animals that live on land drink freshwater, while animals in the ocean drink salt water. The chemicals in their bones can indicate which kind of water they drank. The evidence showed that *Pakicetus* lived in and around freshwater rivers, but *Ambulocetus* probably lived on a swampy ocean coast.

A *Dorudon* fossil lies in an Egyptian desert.

Source: Photograph by Tom Horton, Shanghai American School, *www.flickr.com/photos/further_to_fly/2368052851*

Around the World

Sometime after *Ambulocetus*, whales moved into the ocean full time. Thousands of ancient whale fossils have been found across the world. Fossils of one genus of ancient whale, *Dorudon*, have even been found in North and South Carolina. As the whale fossils get more recent, their hind legs get smaller and smaller. *Dorudon* had a small pelvis and tiny hind legs that weren't strong enough for swimming. In modern whales, these hind legs have disappeared, except for the partially formed pelvis and legs buried in their fat. This has sent scientists who study genes off on a hunt of their own. What kind of genetic mutation could lead to the gradual loss of hind legs?

Scientists have been looking to dolphins to answer this question. Dolphins, which are considered whales by scientists, start to develop legs before they are born. The front legs develop into flippers, but the hind legs stop developing and disappear before birth. Scientists have discovered a number of genes that are important for controlling hind leg development. One gene turns the process on. A second gene keeps it going. For legs to develop completely, the second gene has to stay active. If it turns off too soon, only part of a leg will form.

This multigene system may explain how whales gradually lost their hind legs. The second gene turned on for a shorter and shorter period of time leading to shorter and shorter legs in different whale ancestors. In modern dolphins and whales, the second gene doesn't turn on at all. But the gene is still there, and in very rare cases, it is activated. Occasionally, fishermen or scientists find a dolphin or whale that has incomplete hind legs sticking out of its body!

Back in the Frozen Arctic

The mammalian features and the hidden bones in the bowhead whale only hinted at the amazing journey that whales have taken from land to sea. Scientists have used many types of evidence to piece the story together, but there is still much to learn. Maybe you will be the next scientist to discover more about the evolution of whales.

THE BIG QUESTION

Describe how whales changed as they evolved from land animals to sea animals.

Evidence From the Ancient Whales

"Hunting the Ancient Whales" describes several conclusions (claims) that scientists have drawn while studying whale evolution. In the chart below, fill in the evidence that supports each of their claims.

Claim	Fossil Evidence	Other Evidence
Whales evolved from land animals.	From *Pakicetus*:	From DNA:
Later whales moved into the water.	From *Ambulocetus:*	From soil and chemicals:
Whales gradually lost their hind legs over the course of evolution.	From *Dorudon* and other fossils:	From dolphin genetics:

*Whale Evolution Images**

Pakicetus 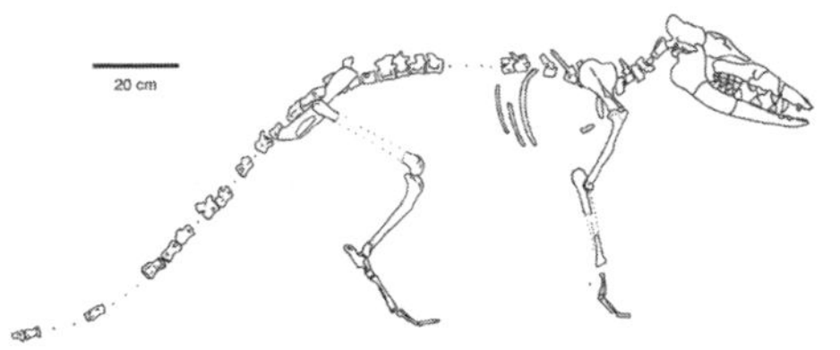	
Maiacetus 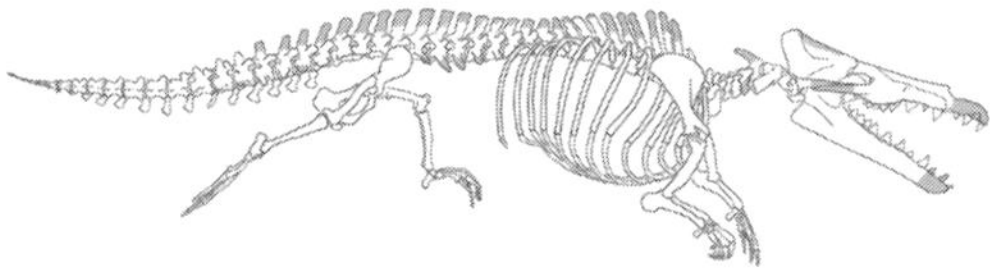	
Dorudon 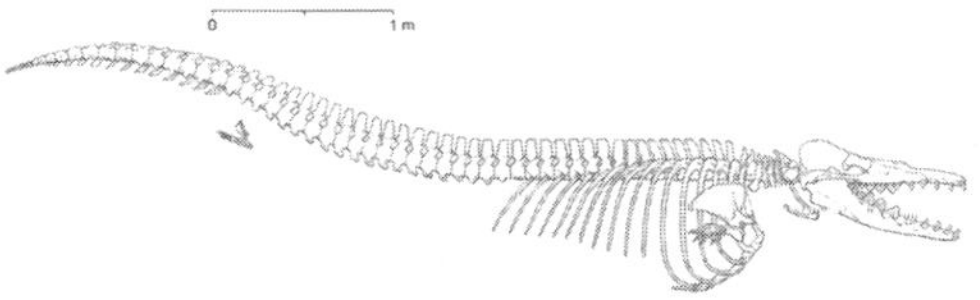	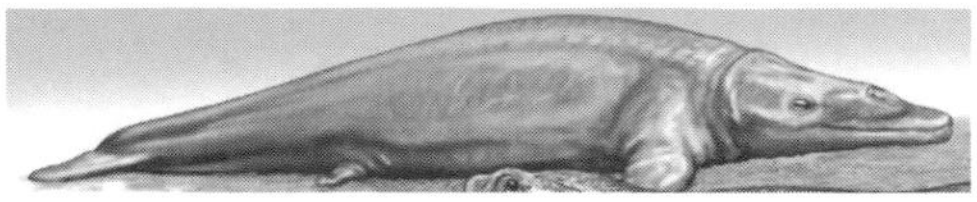
Ambulocetus 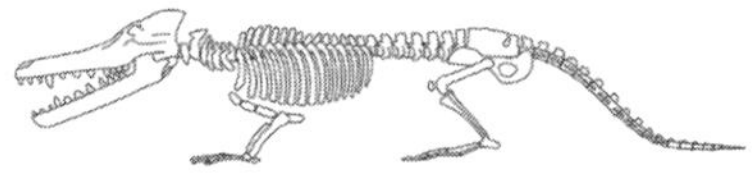	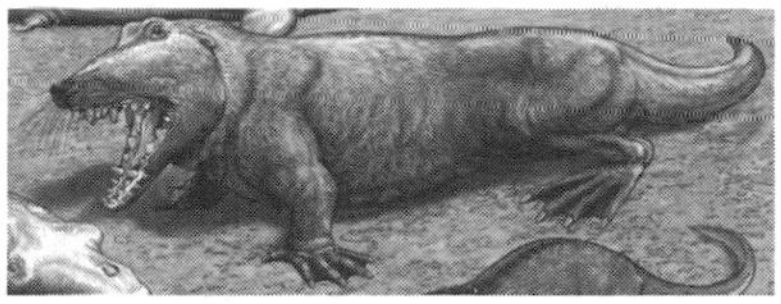
Physeter 	

* Image source information can be found on page 156.

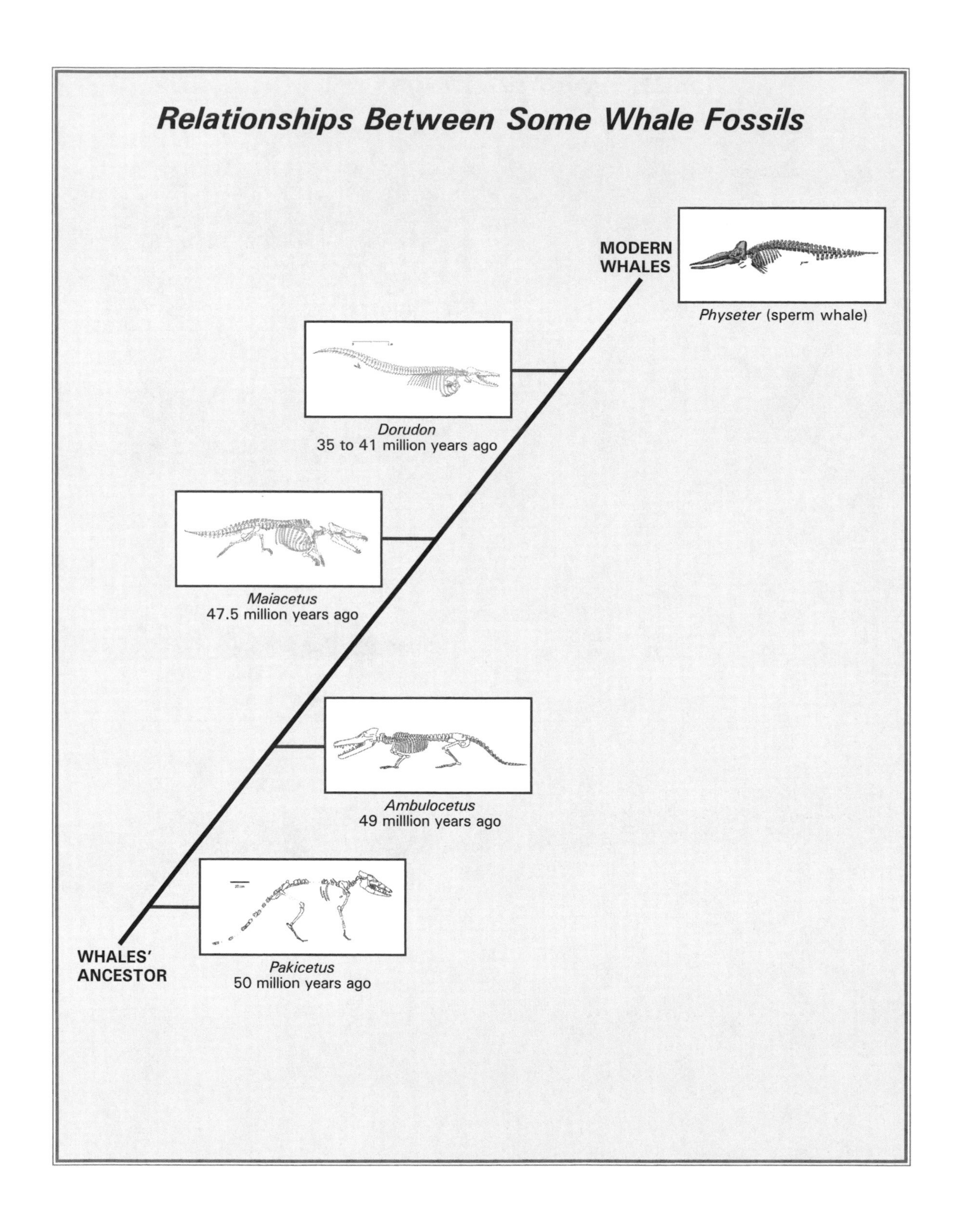
Relationships Between Some Whale Fossils
MODERN
WHALES
Physeter (sperm whale)
Dorudon
35 to 41 million years ago
Maiacetus
47.5 million years ago
Ambulocetus
49 milllion years ago
WHALES'
ANCESTOR
Pakicetus
50 million years ago

Going Batty

Horseshoe Bat Skeleton

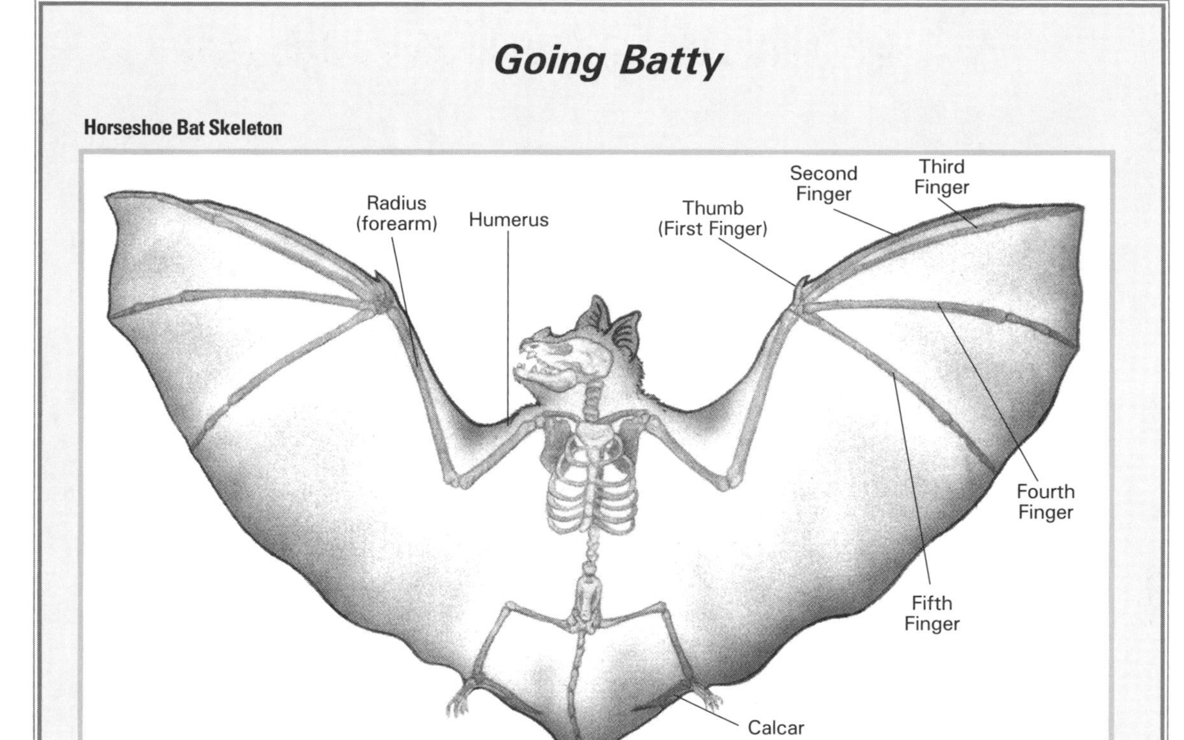

Bats also have an interesting evolutionary history. They are mammals, like humans and whales, but they fly in the air like birds. Bat ancestors would have lived on land. How did they develop into flying mammals? Scientists haven't collected as much information on bat evolution as they have on whale evolution, but they are working on it.

Suppose you were a scientist who wanted to study bat evolution. What kinds of evidence would you look for? What would you hope to find out from each kind of evidence?

(Hint: If you get stuck, think about the types of evidence scientists used to learn about whale evolution.)

Image Sources for Whale Evolution Images

Pakicetus skeleton: J. G. M. Thewissen, *www.neoucom.edu/DEPTS/ANAT/Thewissen/index.html*

Pakicetus drawing: Illustrated by Carl Buell, and taken from *www.neoucom.edu/Depts/Anat/Thewissen/whale_origins/whales/Pakicetid.html*

Maiacetus skeleton: Gingerich, P. D., et al. 2009. New Protocetid Whale from the Middle Eocene of Pakistan: Birth on land, precocial development, and sexual dimorphism. *PLoS ONE* 4 (2): e4366. *www.plosone.org/article/slideshow.action?uri=info:doi/10.1371/journal.pone.0004366&imageURI=info:doi/10.1371/journal.pone.0004366.g001*

Maiacetus drawing: John Klausmeyer and Bonnie Miljour, University of Michigan Museums of Natural History

Dorudon skeleton: Gingerich, P. D., et al. 2009. New Protocetid Whale from the Middle Eocene of Pakistan: Birth on land, precocial development, and sexual dimorphism. *PLoS ONE* 4 (2): e4366. *www.plosone.org/article/slideshow.action?uri=info:doi/10.1371/journal.pone.0004366&imageURI=info:doi/10.1371/journal.pone.0004366.g001*

Dorudon drawing: Reprinted, with permission, from the *Annual Review of Ecology and Systematics,* Volume 33 © 2002 by Annual Reviews. *www.annualreviews.org*

Ambulocetus skeleton: J. G. M. Thewissen, *www.neoucom.edu/DEPTS/ANAT/Thewissen/whale_origins/index.html*

Ambulocetus drawing: Reprinted, with permission, from the *Annual Review of Ecology and Systematics,* Volume 33 © 2002 by Annual Reviews. *www.annualreviews.org*

Physeter skeleton: *http://commons.wikipedia.org/wiki/File:Sperm_whale_skeleton.jpg*

Index

*Page numbers in **boldface** type refer to tables or figures.*

About the Author

Jodi Wheeler-Toppen is a science teacher and children's author. After realizing that her students' reading problems were getting in the way of their science learning, she returned to school to learn more about helping students with reading. She finished her PhD in science education and now teaches adjunct courses at the University of Georgia, with her favorite course being Teaching Science to Students With Special Needs. She lives in Atlanta with her husband and two children.